Appraisal & Evaluation Library

Knowledge Based Systems Volume

July 1991

London: HMSO

Appraisal and Evaluation
Knowledge Based Systems Volume

© **Crown Copyright 1991**

First Published 1991
Applications for reproduction should be made to HMSO

ISBN 0 11 330570 2

For further information regarding this document please contact :-

Strategic Programmes Division
CCTA, Norwich
0603 694706

Foreword

This document is the Knowledge Based Systems volume of the CCTA's Appraisal and Evaluation Library. This Library is intended to aid appraisal and evaluation of data management products and consists of an Overview and Procedures volume, together with supporting technology specific volumes.

The Overview and Procedures volume describes the series and provides a procedure for using the criteria contained in the technology specific volumes in a number of contexts. These include making a strategic selection, evaluation during a feasibility study, and evaluation during the procurement stage of a project. The evaluation procedure is placed into the context of other CCTA procedures, such as those for procurement and evaluation, and methods such as SSADM. It has been written in support of the CCTA Information Systems Guides.

Each technology specific volume provides a hierarchy of criteria that may be used as the basis for the evaluation of products in that technology class. The initial volumes were for Database Management Systems and Application Generators. This volume on Knowledge Based Systems joins the library accompanied by a volume on Text-based Information Management Systems.

This Appraisal and Evaluation Library has been produced to assist organisations to identify the product, or set of products, which best meets their requirements. The procedure and the criteria have developed as technology has changed, and as a result of experience gained from their use. CCTA welcomes comment on, and contributions to, this Library to ensure that it continues to provide maximum benefit.

Appraisal and Evaluation
Knowledge Based Systems Volume

Contents

Introduction page

i	General	7
ii	Scope	11
iii	Criteria	21

Chapter

1	**Functionality**	25
	1.1 Knowledge representation	25
	1.2 Inference	33

2	**Developer interface**	37
	2.1 Editors	37
	2.2 Debug/trace facilities	38
	2.3 Testing aids	38
	2.4 Compiler/interpreter	39
	2.5 Knowledge acquisition	39

3	**Integration**	41
	3.1 Operating system	41
	3.2 Transaction processing systems	41
	3.3 Databases	42
	3.4 Application program interface	43
	3.5 Cooperative processing	43

4	**Efficiency**	45
	4.1 Knowledge representation	45
	4.2 System resource usage	45

5	**Development methodology**	47
	5.1 Methodology	48
	5.2 Compatibility with GEMINI	49

6	**Quality and control**	51
	6.1 Documentation	51
	6.2 Development control	51
	6.3 Audit control	52
	6.4 Quality assurance monitoring	53

Intelligence. Initially they were known as 'Expert Systems' (and are still often referred to as such) as the objective of the early systems was to emulate the performance of a human expert in the performance of some task, e.g. emulating a doctor in the task of diagnosing a patient. The initial attempts at building such systems were characterised by the need for high cost investment in specialised hardware and development tools and, of course, programmers (generally called 'Knowledge Engineers'). This environment proved effective in developing impressive prototypes, but failed when it came to delivering the applications into everyday use and demonstrating business benefit.

As 'Expert Systems' were applied it was realised that the applications for which they were best suited did not require to have as their objective the emulation of a human expert. Instead, substantial benefits could be achieved by capturing and representing small amounts of 'knowledge' and making it widely available. Hence the term Knowledge Based Systems seemed more applicable. This change in objective encouraged the tool developers to deliver their products in conventional data processing environments. The result is the products seen today; providing powerful knowledge representation and inferencing techniques within a conventional data processing environment.

i.3 Audience

The main audience for this document is Information Technology (IT) staff wishing to carry out appraisals or evaluations for soundly based procurement.

This volume will also be of interest to senior IT management considering the introduction of Knowledge Based Systems products and wishing to ensure that such an introduction is carried out professionally, resulting in the selection of the most appropriate product.

It is assumed that the reader has at least a basic understanding of data processing. Knowledge of KBS is not assumed, with introduction ii providing some background information and explanation of the terminology used in this volume.

Because of these assumptions experienced IT staff may find the volume (outwith the Functionality chapter) too descriptive in some parts, but technically simplified in others. Similarly, staff with KBS experience may find the Functionality chapter too descriptive in parts. It should be remembered, however, that the document will be used as a primer by those unfamiliar with the topic, while also serving as a useful reference document for experienced people.

i.4 **Expected uses**

It is expected that the volumes in this Library will be used in several ways. The uses identified in the Overview and Procedures Volume are:

- strategic, business-based, evaluation of products to select a 'standard' product for subsequent organisation-wide use.

- less detailed evaluation of products as an element of a feasibility study.

- full evaluation of products during procurement for a project.

- independent appraisal of a product.

i.5 **Structure of this volume**

This document is in three parts - introductions, the evaluation criteria and annexes.

This is the first introduction. Introduction ii describes the scope of the subject area and explains the terminology. Introduction iii describes the notation used for the criteria, and summarises the main headings.

The bulk of the volume contains the high level criteria and the checklists of detailed technical questions used within the evaluation model to assess and rank KBS products. The questions can be used as an aide-memoire when gathering information about products.

Annex A contains a hierarchy chart of the subject matter in this volume. This chart may be used as a

Appraisal and Evaluation Library
Knowledge Based Systems Volume

default or as the basis for a hierarchy chart which best meets the needs of the project or organisation. Annex B contains a list of references.

i.6 Outline of the procedure

The evaluation process comprises 7 stages which are described in the Overview and Procedures volume.

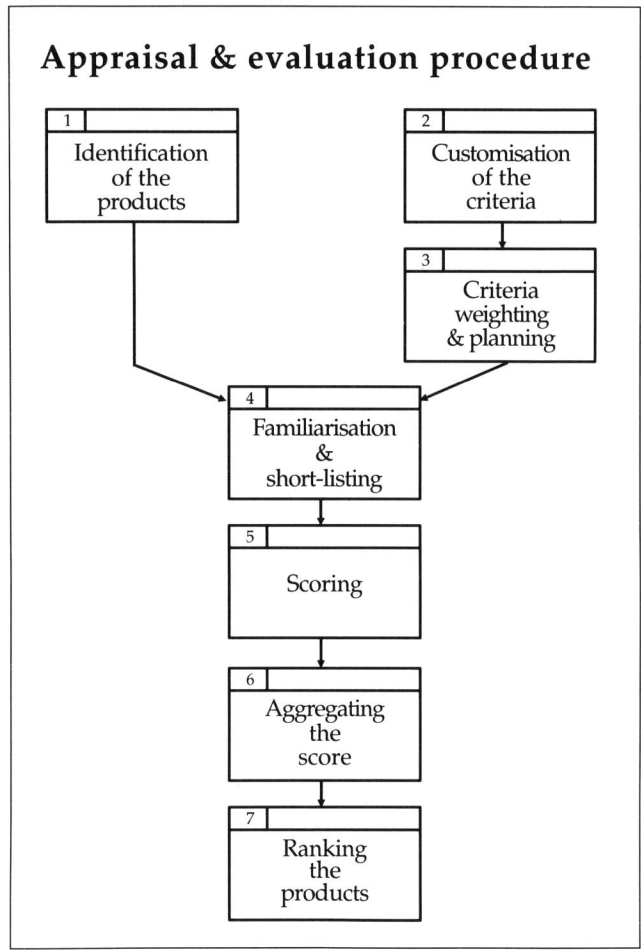

i.7 History

This is the first version of this volume and was written for CCTA by Phil McKell then of the Turing Institute, Glasgow.

ii Scope

ii.1 Scope of this volume

The evaluation criteria in this volume relate to KBS software products appropriate for the construction of multi-user applications, both stand-alone KBS applications and those in which the KBS is but one component. The criteria are, where possible, consistent with those of the companion volume on Application Generators which, together with the evaluation method described in the Overview and Procedure volume, have been successfully applied in both strategical and tactical procurements. Extensions to this volume will be required as demand for specialised KBS such as real-time KBS, parallel architecture KBS and distributed KBS grows.

ii.2 KBS Definition

The KBS arena can be very confusing for a beginner because of the multiplicity and contradictory nature of the terminology used by the product vendors and in the press. The term 'Knowledge Based System' is used to refer to both applications and tools, and is used synonymously with the term 'Expert System'. Although there is no universally accepted definition for either of these terms, a working definition of an 'Expert System' is:

> ' a computer program that embodies the knowledge of a specialist in a particular domain, which can advise non-specialists with respect to that domain and offer explanations for the advice given'.

ii.3 KBS tools

This section describes the types of KBS tools available to build these systems. A KBS tool is software package specially designed to support the construction of KBS applications. Such a tool usually provides a language for representing knowledge, inference techniques for processing the knowledge, and an environment for developing the application (the 'development environment').

Architecture

Although the family of KBS tools is quite diverse, most tools are based on the 'production system' architecture of early expert systems, which is shown below in Figure 1:

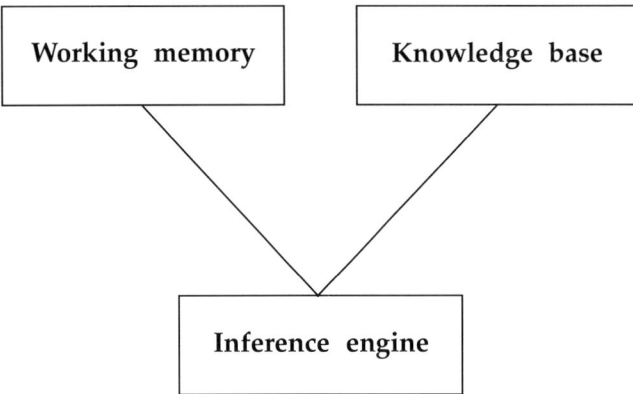

Figure 1: A Production System Architecture

The main features of this architecture are:

- The **knowledge base** contains the problem solving (i.e. the expert's) knowledge. Traditionally, this (knowledge) was represented solely using IF-THEN rules. Nowadays, however, it is common for both *object-oriented* techniques and rules to be provided in a KBS tool (see explanation of object orientation on page 18).

- The **inference engine** contains (control) knowledge of how to apply the knowledge within the knowledge base to solve a given problem. This is generally in the form of one or more reasoning paradigms (see later) such as forward and backward chaining, opportunistic reasoning, and hypothetical reasoning. It is provided as part of a KBS tool.

- **Working memory** is the storage into which data describing the current (problem) situation is placed. The Inference engine applies the knowledge (such as rules) within the knowledge base to this in order to derive a solution.

A **blackboard system** is an extension of the production system architecture. In this, multiple experts (inference engines) cooperate to solve a problem, which is described on a global 'blackboard'. A 'referee' dynamically selects the expert most able to further the solution of the problem.

Classification

Artificial Intelligence (AI) Languages: The two most well-known of these are LISP and Prolog. Lisp (LISt Processing) was designed in the mid-50's to support programming by symbol manipulation as opposed to calculation, and is popular in the USA. Prolog is based on predicate logic and is popular in the UK and Europe. Both languages are flexible but require skilled staff to develop applications.

A **shell** consists of an empty knowledge base, a high level language for representing the knowledge (normally rules), a (typically backward chaining) inference engine, and a user interface for entering and consulting with the knowledge. It is relatively simple to use but restricted in the range of applications to which it can be applied.

A **hybrid shell** is a more advanced shell which offers both backward and forward chaining, and the ability to deal with structured data (objects). It may also offer:

- advanced facilities for developing both character and graphical user interfaces.

- interfaces to common databases.

- portability between conventional platforms such as mainframes and PCs.

Such shells combine ease-of-use with powerful knowledge representation and inferencing techniques. They provide facilities which increasingly rival those offered by toolkits.

Toolkits combine powerful reasoning strategies with an underlying language, which is typically LISP (although Prolog and Poplog are also popular). They do not always integrate easily into a data processing environment and may require highly skilled staff to exploit their full capabilities.

- **Best First:** this is an extension of the breadth-first strategy which guarantees to find the best possible solution but without having to visit every possible node in the search space. It achieves this by evaluating every node at the deepest level of the tree and selecting the best one (in terms of moving closer to the solution) to expand.

Reasoning

Reasoning, or 'chaining' as it is typically referred to in KBS tools, is the process by which knowledge is applied to move closer to the solution to a problem. The most common reasoning techniques are: Forward Chaining, Backward Chaining, Opportunistic Reasoning, Hypothetical Reasoning, Nonmonotonic. Reasoning and Demons.

- In **forward chaining**, sometimes called 'data-driven' searching begins with a number of facts (in the initial states) and a set of legal moves (rules) for generating new facts. These (new) facts are themselves used by the rules to generate even more new facts. This process continues until a path to a solution is reached.

  Forward chaining is appropriate to problems in which all or most of the data are given the initial states. It is also appropriate when there is a large number of potential goals, but only a few ways to use the facts, or it is difficult to form a goal or hypothesis.

- Whereas in forward chaining, the process is one of inferring new facts from the initial facts, in **backward chaining**, the hypothesis is that a fact is true (the goal) and then try to prove it. That is, we try to find the facts that could be used to generate the goal and determine what conditions must be true to use them. These conditions become the new goals, or sub goals, for the search. The process continues until it works back to the facts of the initial state given for the problem. For this reason, backward chaining is sometimes called 'goal driven' reasoning.

  Backward chaining is appropriate if a goal can easily be formulated. Additionally, it may be used if there is a large number of rules that match the facts of the problem, or if problem data are not given but must be acquired by the problem solver.

- **Opportunistic reasoning** is where the most appropriate method of applying knowledge at any point in time is chosen dynamically. This choice is generally between backward and forward chaining.

- **Hypothetical reasoning** supports the exploration of multiple hypothetical situations, referred to as *contexts*, *states*, *viewpoints*, or *worlds*. It is often supported by a truth maintenance (nonmonotonic reasoning) system which eliminates any contradictory states (see below).

- In **nonmonotonic reasoning**, or *truth maintenance* as it is commonly known, when something that was previously thought to be true (for example, *the switch is closed*) is later shown to be false, all conclusions that depend on that fact must be reexamined and possibly withdrawn as well (for example, *the light bulb is on*).

- A **demon** (an interrupt) is a piece of code which is executed whenever the condition it is monitoring is satisfied. For example, we may wish the system to perform type checks or run consistency tests whenever a certain datum is changed.

Object-orientation

The object-oriented paradigm is an approach to modelling which builds on the ideas of abstract (real world) objects, encapsulation and class inheritance.

- An **object** contains both data, represented by attributes, and processing, represented by methods. The 'object' behaves (performs a task) in response to receiving a message that it understands.

- A **method** is an internally coded procedure which implements (part of) the functionality of an object. It is actioned when the object receives a specific message.

- **Encapsulation**, also known as 'information hiding', ensures that the internal structure of an object is invisible to all other objects. Encapsulation isolates the object data and methods from the outside world. All communication between objects is in the form of messages.

- A **message** is the mechanism by which one object communicates with another to force the execution of a method.

- A **class** is used to define common attributes and methods for a group of objects. The 'class object' can be considered as a parent of the (child) objects it relates to. A child may have more than one parent.

- **Inheritance** is the hierarchic mechanism by which 'child' objects exhibit behaviour and properties defined for their forebears.

General

This section provides the meaning for terms, not covered above, that are used within the evaluation criteria and/or are in common use within the KBS field.

- The **agenda** is a list of the tasks which the inference engine of a KBS tool is to perform.

- **Artificial Intelligence** is a subfield of computer science, which is concerned with the study of intelligent problem solving behaviour and, in particular, how it can be emulated on a machine.

- **Backtracking** is the process by which a search strategy on reaching a 'false' state backs up the search tree to a previous decision point. An alternative path is then chosen and explored.

- An **expert system** is a computer program that embodies the problem-solving knowledge of a human expert in some specialised domain. The term is commonly used interchangeably with 'KBS'.

- A **frame** derives from AI and is used synonymously with 'object'. However a frame differs from an object in that it has slots as opposed to attributes and 'procedural attachment' as opposed to 'methods'. The difference in processing being that in the 'procedure attachment' the procedure is attached to slots (attributes) and invoked by accessing the slot, whereas with methods the procedure is invoked by sending a message to the object.

- A **heuristic** is a technique (such as a rule of thumb) which determines what alternative solutions to explore in the problem space.

- **Inductive reasoning** is the method of using a set of examples to extract generalised rules which can be applied in all situations.

- **Knowledge** is a combination of data structures and rules/procedures, that if represented and used adequately, will lead to intelligent behaviour.

- **Knowledge acquisition** is the process by which the knowledge to be represented within a KBS is identified, elicited from its source and represented, preferably in an implementation-independent formalism. It is composed of a number of subprocesses: the elicitation of the knowledge from the expert or other source such as rule books, the analysis of the data produced, and the organisation of the concepts identified into a conceptual model of the domain knowledge.

- **Knowledge engineering** is the generic term given to the process of designing and building KBS.

- The **Rete algorithm** is an algorithm which eliminates unnecessary work during pattern matching in rules. When rules are first loaded into the system, they are compiled into a set of features to be checked, such as tests on the values of attributes. Similar tests in different rules are shared, to eliminate duplicate checking. The result is a tree-structured sorting network that efficiently performs the matching process.

- A **rule** is a construct for expressing knowledge. It relates two statements A and B as follows: IF A then B. That is, when we know A is true, we can conclude B to be true.

- **Symbolic processing** is the manipulation of symbols, (representing the knowledge) as opposed to numbers, to solve problems.

iii Criteria

iii.1 Notation

The criteria in this document are structured as a hierarchy, this is illustrated in the Annex A.

The text is in three classes:

- the main discussion of the criteria - it is primarily this text that should be customised for particular projects against which weights are assigned and allotted. To obtain an overview of the criteria this text can be read in isolation. This is printed in 10 point Palatino typeface (ie the one used to print this volume) alongside a numbered heading in bold type, as in the top paragraph on this page. Where the criteria covers a large subject area it is divided into sub-criteria. This is printed in the normal typeface with an unnumbered side heading in the same typeface (ie not bolded).

- detailed discussion of the criteria or sub-criteria - this level is required for information gathering. This is also in the 10 point Palatino typeface, it does not have a heading.

- *the supporting questions associated with the criteria or sub-criteria - these are in italics, as this example.*

iii.2 Summary of the criteria

The hierarchy of evaluation criteria against which KBS tools can be scored is summarised below and elaborated in the chapters that follow. A diagrammatic representation of the hierarchy adopted within this volume appears as an annex. It will, of course, be necessary to construct a hierarchy applicable to the needs of the project or organisation, which may be different to the one we have illustrated. All changes made should be justifiable.

The top level criteria are:

- Functionality - the knowledge representation techniques and inference methods provided.

- Developer Interface - the facilities available to the developer and the skills required for application development using the product.

- Integration - degree to which the product integrates with other software.

- Efficiency - development and run-time machine resource usage.

- Development methodology - degree to which the tool supports any of the emerging methodologies for developing knowledge based systems.

- Quality and Control - the facilities available for, and the degree to which it is possible to ensure, high quality software engineering.

- Portability - the range of hardware platforms and operating systems on which the product will run and the extent to which the development environment or individual applications are transferable.

- End User Interface - the facilities available for developing end-user interfaces.

- Security - protection of and control of access to data and knowledge and to the tool itself, especially the development environment.

- Product Credibility - status of the product and supplier and degree of support, release strategy, training.

- Project Specific criteria - other than the above.

- Costs -assessment of direct and ancillary costs; hardware, software, personnel, maintenance, training, development and operation.

It is expected that these criteria, with the exception of costs, will be weighted and scored as set out in the Overview and Procedures volume of this library. The cost information will be required as an element of the selection procedure, or to exclude products for detailed consideration when they exceed planned budgets or cost ceilings.

Note that the criteria are not intended to form a tutorial on the subject under consideration. There is a wide range of published material available, including several reports and papers from CCTA. Please contact CCTA regarding the availability of appropriate documents.

iii.3 Questions

This volume consists of a discussion of each of the above criteria together with relevant detailed questions. The questions should be used to facilitate familiarisation with a product before attempting to allocate scores against evaluation criteria. Not all questions are relevant to all products, or projects, and they should be used selectively.

Experience has shown that little will be gained by having the vendor provide written answers to the questions. Only by asking probing questions can the evaluation team fully elicit the limits of the capabilities of the products. The best value will be obtained by attempting to answer questions after inspection of technical documentation and attending demonstrations, and talking to current users and visiting reference sites.

Appraisal and Evaluation Library
Knowledge Based Systems Volume

1 Functionality

The functionality of a KBS tool is measured by the combination of its knowledge representation and inference facilities. It is the combination of these facilities and the high level techniques that they provide for representing problem solving knowledge which differentiates Knowledge Based System tools from other software development tools, such as Application Generators. Experience over the last ten years or so has shown that in order for a tool to be both flexible enough to handle a wide range of problems and powerful enough to solve difficult problems it requires a mixture of knowledge representation paradigms. In particular that of rules, objects, and mixed inferencing.

1.1 Knowledge representation

The most common techniques for representing data (or knowledge) in KBS are those of rules and objects. Rules represent the heuristics employed by the expert to solve the problem, while objects are used to structure the domain knowledge, both physical and/or conceptual entities.

1.1.1 Objects

An object consists of both data (normally called attributes) and procedures (normally called methods). It hides its data and the implementation details of its procedures from the outside world behind an interface (or protocol) defined by the names it has given to its procedures, these are the messages to which the object will respond.

Attributes

Attributes describe the features of an object (like a tuple in a relational database). In order to provide flexibility a wide range of attribute types should be supported, including: logical (boolean), numeric, string, multi-valued, bitmap, symbolic and other (where other could be array, list, time, interval, rectangle). In a true object-oriented system access to the values of these attributes is achieved only by sending a message to the method which implements the required behaviour, thus enforcing the idea of encapsulation. However in knowledge based applications it is often possible able to access the attributes directly from within rules, to compare attributes within the same object or attributes within different objects (similar to a relational join).

- the use of the familiar 'IF condition THEN action' syntax.

- the ability to define problem specific relations within both the condition and action parts.

- making the rule look like 'real english' by using 'noise words' - words ignored by the compiler, such as 'the', 'a', 'some' and 'all'.

Ideally, there should be no restrictions on the number of conditions that can be placed on the left hand side, (LHS) or actions (conclusions) on the right hand side (RHS), of a rule. The rule syntax should also permit the placement of calls to built-in and external functions.

How readable are the rules that can be produced?

Can problem-specific relations be defined?

Can noise-words be defined?

What connectors are provided (AND/OR/NOT)?

Are there any limits imposed on the number of conditions in the Left Hand Side (LHS) or conclusions on the Right Hand Side (RHS) of a rule?

Can rules have an ELSE clause?

Can calls to built-in and external functions be made in either or both the LHS and the RHS?

How extensive are the built-in functions (Maths, Statistical, String Handling, Financial)?

In addition to the numerical and string handling functions a KBS tool should also provide the rule language with the ability to invoke the inference engine via calls such as 'forwardchain', 'backwardchain', and modify the current control strategy by modifying attributes of the inference engine.

Are there facilities to modify the current control strategy?

Can these be used as part of a 'standard rule'?

Can the inference engine be invoked from the RHS of a rule?

Chapter 1
Functionality

If multiple conclusions are allowed, can all of them invoke the inference engine?

Can the current control strategy be modified from the RHS of a rule?

Pattern matching (expressibility)

The advent of hybrid systems, offering both rules and objects, has created the requirement that the rule language should have proper access to the object hierarchy. In particular it should provide flexibility in expressing patterns of object-attribute relationships against which the rules can match, with the use of variables allowing the expression of generic patterns. Once again flexibility is desirable with the ability to use a variable in place of an object, an attribute or a value. This can be a very powerful construct when developing knowledge based applications that reason with data stored in a database.

Can variables replace objects, attributes, values, in expressing conditions in rules which match against the object hierarchy?

Are any 'meta-constructs' provided for pattern-matching, such as:

- *Find one instance?*

- *Find all instances?*

- *Count the instances?*

- *Create an instance?*

Objects matched on the LHS should be accessible in the RHS of the rule in order that the required actions can be performed on the matched data. If the objects matched are collected together in a list and passed to the RHS of the rule then list manipulation functions, such as first_element, last_element, rest_of_list, any_element, are required in order to process this list. If a binding of the rule is created for each object matched then the ability to compare two members of the same class of object within the same rule is required to express statements requiring inner-class comparison, such 'give me the biggest X'. This is normally provided by way of a 'qualifier', e.g. 'if

1.1.4 Uncertainty

In real-life, outcomes may not always be absolutely clear. For example, the fact that an outgoing flight has been delayed is 'somewhat' an indicator that the plane has mechanical problems. However, it is also 'somewhat' an indicator that the airport is busy. In KBS terminology, we would say that both facts are *uncertain*.

The most common approaches for managing uncertainty in KBS tools are *Bayesian probability* and *certainty factors*:

- **Bayesian probability**: in this (formally-based) approach the probabilities of previously known results are used to calculate more complex probabilities, using the following theorem:

$$p(Hi|E) = \frac{P(E|Hi) * P(Hi)}{\sum_{k=1}^{n} (P(E|Hk) * P(Hk))}$$

where:

P(Hi|E) is the probability that Hi is true given evidence E.
P(Hi) is the probability that Hi is true overall.
P(E|Hi) is the probability of observing evidence E when Hi is true.
n is the number of possible hypotheses.

and all relationships between evidence and hypotheses, or P(E|Hk), are independent.

Bayesian reasoning must be used with caution. First, complete and up-to-date probabilities are necessary if its conclusions are to be correct. In many domains, this is not always possible. Second, all probability relationships must be rebuilt when any new relationship of hypothesis to evidence is discovered. Finally, it is questionable whether human experts use the Bayesian model in problem solving.

- **Certainty factors**: this approach makes simple assumptions for creating confidence measures and uses simple rules for combining them. For example, the certainty factor (CF) of the conjunction (ANDing) of two premises might be the minimum factor:

    CF(P1(0.6) and P(0.2)) = MIN(0.6, 0.2) = 0.2

    whereas for a disjunction (ORing), the maximum factor might be used:

    CF(P1(0.8) and P(0.3)) = MAX(0.8, 0.3) = 0.3

    Other calculations describe how to combine multiple CFs when two or more rules support the same result R.

    Certainty theory, although it has a formal basis, has received criticism for being excessively ad hoc. Many shells only provide guidance to users on establishing their own certainty factors.

What facilities are provided for handling uncertainty (certainty factors, Bayesian probability)?

Can these be switched on/off within a knowledge base or individual rulesets?

1.2 Inference

The inference engine of the tool is central to its knowledge representation powers and is the main difference between hybrid KBS Tools and object oriented tools like C++ and Smalltalk. It determines the manner in which the knowledge in the knowledge base is applied to solve the problem at hand. The Inference component of a KBS tool can be examined under two categories: Reasoning and Control.

1.2.1 Reasoning

There are four common reasoning strategies: Backward Chaining, Forward Chaining, Opportunistic Reasoning, Hypothetical Reasoning. Of these both Backward Chaining and Forward Chaining are essential ingredients of a flexible KBS tool. With respect to Opportunistic reasoning most tools will offer a basic version known as 'mixed inferencing' where the inference engine can switch from backward to forward chaining freely. The more advanced is found only in those tools supporting the Blackboard paradigm.

Hypothetical Reasoning supports the simultaneous exploration of a number of hypothetical situations, referred to as *states*, contexts, viewpoints, or worlds. In the advanced toolkits, it is supported by a Truth Maintenance (TM) system which watches for any contradictions which may arise, and eliminates the offending state when one if found. In other tools, Hypothetical Reasoning is not supported by TM, instead multiple paths are maintained. A best-first search strategy selecting the path with the best rating for expansion at each step in the problem solving process.

What reasoning modes are provided by the tool? (Backward Chaining, Forward Chaining, Opportunistic Reasoning, Hypothetical Reasoning)?

Can opportunistic reasoning be set to allow only backward chaining or forward chaining if required?

Is hypothetical reasoning supported by a truth maintenance system or by an implementation of the best-first (A*)algorithm?

1.2.2 Control

This is concerned with the facilities offered by the tool for deciding 'what to do next'. With 3GL's this is easily answered as 'execute the next statement unless in a loop or GOTO'; with KBS tools the answer is not quite so simple. The task of identifying the control facilities offered by a KBS tool is eased if based it on a three step model of the inference cycle:

> Step 1 - the pieces of knowledge (rules or rulesets) that can fire are identified. This will be achieved either by looking at the LHS of the rules (or activation condition of the ruleset) if the reasoning strategy employed is forward chaining. If it is Backward Chaining then this will be achieved by examining the RHS of the rules (or the Goal attribute of the rulesets).
>
> Step2 - conflict resolution - the selection of the rule (or ruleset) to fire. This will be achieved by

prioritising the rules (or rulesets) and defining a strategy for comparing the priorities and selecting the most applicable piece of knowledge (rule or ruleset).

Step 3 - the rule (or ruleset) is fired.

Construction of conflict sets	The flexibility here depends on the pattern matching ability of the rules (see earlier section) and the reasoning strategy employed.

What is the pattern matching ability of the rules?

What reasoning strategies can be invoked (backward or forward chaining)?

Conflict resolution	The flexibility here is determined by the tool-supplied attributes upon which the elements of the conflict set can be ordered. It is through the use of priorities that the various search strategies (depth-first, breadth-first, best-first) can be implemented. Within any of the above reasoning strategies various search strategies can be applied, together with various levels of control.

What system attributes are supplied for determining the priority of rules: (placement in code, recency, certainty, weighting)?

Can priorities be assigned to:

- *rules?*
- *objects?*
- *rulesets?*

In what ways can the priorities be compared to determine the next action to take?

Are there separate conflict sets for rules and rulesets?

Can the priority value be calculated dynamically?

Can the function calculating the priority access the global database (objects)?

Truth maintenance	A computational technique for control of nonmonotonic inference. Deduced information is labelled with the assumption on which it is based, if any of these

assumptions become invalid then the information is retracted. There are two kinds of TM: assumption based stores the context in which the assumption applies, justification-based does not store the context.

Is a truth maintenance system supplied? If so what kind?

2 Developer interface

The emphasis in building knowledge based applications is on the rapid development of prototypes to aid further extraction of knowledge from the expert. This requires easy-to-use graphical editors and browsers. Additionally some tools may provide facilities to aid in the acquisition of knowledge from the expert such as inductive front ends. The user interface is normally central to the success of the application, unless the knowledge base is to be embedded within another system, and therefore the provision of tools to facilitate the construction of both character based (important in a mainframe environment) and graphical based (important to take advantage of powerful workstations and client/server architecture) user interfaces is important.

2.1 Editors

The editing facilities provided can range from simple text editors to graphical editors. A tool would preferably have both a powerful text editor and a graphical editor. The latter is useful when learning the tool and for speedy prototyping, but can become cumbersome when full-scale development is required. At this point a powerful text editor is helpful.

Graphic editors are most powerful when they show both the structure of the rules and the object hierarchy within the knowledge base, and allow the editing of rules or objects by selecting the appropriate entity. This greatly facilitates the process of maintenance by providing an overview of the structure of the knowledge contained within the system.
The ability to customise the editors, such as the colours used to represent rules and objects, to suit the developers own preferences, and the speed at which the editor interacts with the developer are both important.

What editing facilities are provided with the tool?

Are graphical object and rule browsers provided?

Can multiple windows, viewing different parts of the KB, be opened simultaneously?

Can editing facilities be customised to suit personal preferences?

Are they fast enough?

Are rule, object (syntax) templates provided?

2.2 Debug/trace facilities   The provision of a tree which shows the paths traced during a consultation of the KBS is very useful for debugging, as are breakpoints and stepping utilities. The ability to attach breakpoints to attributes, rules, and rulesets greatly facilitates debugging.

Can breakpoints be set dynamically?

Are breakpoints allowed on rules and object attributes?

Are stepping facilities together with history supplied?

Is a tracing facility supplied?

2.3 Testing aids   A test case library facility allows standard consultations to be created and tested against whenever changes are made to the knowledge base. The facility should allow the automatic saving of cases and the running of multiple cases for testing.

Are history and replay facilities provided?

What facilities are provided for verifying/validating the knowledge bases created?

Is there a test case library facility?

Is there a facility for performing regression testing?

Some KBS tools provide facilities for identifying possible errors within rulesets. Tests that can be provided include:

- testing for unused consequences of rules.

- conditions of rules that cannot be satisfied (although this could be due to the fact that the condition is established by another ruleset).

- identifying circular reasoning.

Are any facilities provided for identifying errors within rulesets such as unused consequences, unreachable rule conditions, circular reasoning?

Can you edit during testing and then continue the test?

2.4 Compiler/interpreter

KBS tools normally provide an interpreter and/or a compiler for the development environment. An interpreter is useful during development for experimenting with new constructs. A compiler can produce efficient code when development is finished. The tool should also allow a run-time only version of a knowledge base to be created. This is the development environment with the editing and compilation features removed, thus protecting the knowledge base from amendment by the user. It can greatly reduces the cost of distributing a KBS application.

Does the development environment provide an interpreter and/or compiler?

Are run-time only versions available?

2.5 Knowledge acquisition

Techniques exist to help the developer extract the knowledge from a human expert. A substantial amount of research has been carried out in this area and is still progressing, and such facilities are now beginning to be supported by KBS tools. A useful aid in building knowledge based systems is an inductive front end, which allows rules to be extracted from the expert via examples. It can be used to 'kick-start' the extraction of rules from the expert.

Are any knowledge acquisition aids provided?

Is the code produced by the knowledge acquisition tool directly available to or amendable by the knowledge engineer?

Appraisal and Evaluation Library
Knowledge Based Systems Volume

Chapter 2
Developer interface

3 Integration

The issues to be considered here are similar to those for Application Generators. The manner in which the tool interfaces with the operating system, transaction processor, DBMS, 3GL's and 4GL's should be investigated here. Also the general move to a client/server architecture is prompting the appearance of facilities to support Cooperative Processing in some KBS tools.

3.1 Operating system

The tool should interact fully with the operating system on which it is running. This implies that it should provide the developer with full access to the facilities available for managing the system resources required by the application. This is of particular importance on workstation platforms. On a mainframe platform the interface with the Transaction Processing Monitor is more likely to be important. Investigation should be made into the impact that the tool will have on system resources (see chapter of Efficiency).

Can the tool communicate with the host operating system via system calls?

Are there any restrictions on how system calls are invoked or data passed within a call?

3.2 Transaction processing systems

If a KBS tools is to be used within a mainframe environment it is essential that it can be fully integrated with the Transaction Processor (TP) monitor. This implies that the TP monitor and KBS tool can communicate in such a way that the TP monitor controls the interaction with both the users and any databases accessed.

Can the KBS tool act as a server to the TP monitor (as a DBMS server does)?

Is the KBS tool capable of multi-tasking? Does this apply to applications as well?

Are both the KBS tool and the applications developed with it fully re-entrant?

Does the TP monitor handle the storing of intermediate results for multi-phase transactions?

3.3 Databases

The interface between the KBS tool and existing databases is of major importance. At present considerable effort can be spent in providing the KB with access to the required data. However, with the increasing move to open systems and the standardisation this brings the situation is improving, with SQL emerging as the industry standard for accessing databases.

Applications developed with the tool should be able to interact with the DBMS in a run-time environment to change, create, and retrieve DB entries with DB retrievals able to create and update objects in the KB. If the tool supplies a satisfactory interface to the DBMS to be used then much effort will be saved in the development effort. If no interface exists then some estimate should be made as to the effort required to construct the interface.

What DBMS and file formats are interfaces provided for?

Are the Data Manipulation Language (DML) statements coded explicitly within the Knowledge Base or are they generated automatically? (ie the KBS developer treats all objects as if they were within the KBS application).

How are data values passed between the Knowledge Base and the DBMS?

What is the maximum number of files, records, fields, characters beyond which the interface capabilities cannot be guaranteed?

Can one or more interface be run on the same processor simultaneously, and can a particular application use both interfaces?

What happens to RDBMS NULL values?

How is the KBS value UNKNOWN (and/or unevaluated) stored?

If the KBS tool does not provide an interface to a particular file format or DBMS, what effort is required to develop one?

SQL is becoming a de facto standard for communicating with databases. If the KBS tool supplies an SQL interface it is important to determine a number of facts.

Does the tool adhere to the current ANSI standard for SQL?

If enhancements have been made to the standard, what are they?

Can all features of SQL be utilised from the interface provided by the KBS tool?

If some desired SQL-access strategies are not handled by the interface can the SQL be hard-coded?

Are SQL calls dynamically bound, or can they be precompiled for better performance?

3.4 Application program interface

The tool should provide an interface to external programs which allows both the embedding of calls to external programs from within the knowledge base, and the embedding of calls to the knowledge base from within external programs. Calling another processing routine implies that the processing control returns to the calling routine. In some cases control passes from one routine to the next (chaining). It is important to investigate which is available. Attention should be paid as to how data is transferred during this process.

Can a 3GL or 4GL be called from within the KB? If so, which?

Can a KB be called from a 3GL or a 4GL? If so, which?

Can a 3GL or 4GL be chained to from within the KB? If so, which?

Can a KB be chained to from a 3GL or a 4GL? If so, which?

How is data passed in all cases?

3.5 Cooperative processing

The process of dividing an application between two machines, in particular between a workstation and a mainframe is assuming increasing importance. Many of the KBS tools now claim to support this facility.

How is cooperative processing accomplished?

Between what platforms can cooperative processing take place?

What is required on the two machines?

4 Efficiency

There is a natural concern that applications built using KBS tools will be susceptible to performance worries with regards to efficiency of CPU and resource usage, which may result in a slow response time for online systems. In addition to determining if any benchmarks exist for the product, questions can be asked with regards to how certain aspects of the tool's functionality are implemented.

4.1 Knowledge representation

The manner in which its knowledge representation facilities are implemented, and in particular its pattern-matching (if any) and object hierarchy (if any). Also the presence of many demons can slow down processing.

Is the RETE algorithm used? If not how are the rules matched to data?

Is the manner in which rules are grouped likely to cause performance problems?

How is the object hierarchy implemented?

Is it possible to establish how resource intensive this facility is?

Is it possible to create object instantiations while running the application or must a maximum be allowed for throughout execution?

Are rules/methods interpreted or complied?

What impact on system performance does the presence of many demons have?

4.2 System resource usage

It is important to be aware that the measurement of the efficiency of a KBS tool is not simply a matter of measuring CPU resources used. Memory usage is also important if the tool is not effectively multithreaded, and so requires significant storage resources. This is of particular importance in a multi-user and/or multi-tasking environment.

Can the knowledge base be compiled into object code to improve CPU speed and reduce memory requirements?

Does the tool provide run-time optimisation facilities?

How efficiently has the product been optimised for the target hardware and systems software environment?

Is there a minimum overhead impact per application?

How much memory will each user of the application require?

Does the overhead lessen or grow with additional applications?

Is there a maximum number of applications that can be running at the same time on one mainframe processor?

How much memory does the inference engine require?

Can knowledge bases be shared across address spaces?

Does the run-time environment allocate its own address space?

5 Development methodology

The increasingly wide use of KBS, both as a component of a traditional application and as a standalone KBS, is generating a demand for a methodology to guide the KBS development process. In the companion volumes in this series (Application Generators and Database Management Systems) this chapter offers guidelines on how to determine 1) whether the advice given by the supplier with regards to application development is compatible with the approach advocated by SSADM and 2) if the tool offers facilities which supports the development of applications in accordance with SSADM guidelines. At present, much effort is being devoted to the development of a similar methodology to support the development of KBS applications. Although a few have emerged and have been applied (e.g. KADS, STAGES), none has yet reached the maturity and widespread use achieved by SSADM. This gap is being tackled by GEMINI (General Expert Systems Methodology Initiative), a joint government and industry venture.

While not yet complete, and therefore not available for adoption, work to date has established a number of things. First that the prevailing 'rapid prototyping' approach to the development of KBS applications is not a suitable methodology on which to base the development of future systems. Instead it is a technique which can be put to good use within one or more of the phases of the overall development lifecycle. Basically the view from efforts to date is that the lifecycle for KBS development is very similar to that of conventional information system development. The differences occurring in the constituent activities of these phases, the rules and guidelines on how these activities can be ordered, and the tools and techniques employed to support the development of a KBS application. These include a strong emphasis on project reviews, and the continual assessment, in the early phases, of the feasibility of the application.

Given the above it is hoped that once GEMINI becomes established then similar conformance guidelines to those provided for SSADM, can be generated. However for now, some pointers can be provided to help determine if the supplier is able to provide sensible advice with regards to developing KBS applications with their tool, together with a spiral model of software development.

5.1 Methodology

Does the supplier advocate a rapid prototyping approach, utilising the powerful development facilities offered by their product, or (as they should) do they emphasise that rapid prototyping should only take place within an overall lifecycle process?

Does the vendor offer their own methodology or recommend one? If so, does their product provide any help in using it?

Can the supplier offer help in determining the feasibility of the intended application?

Is emphasis placed on producing a logical design of the knowledge to be represented, which is independent of the physical realisation of the system?

Is the role of prototyping given as that of the main system development technique or (as recommended) as a technique which may be used in both the analysis and the design phases, but with strict criteria to guide and control the prototyping?

What advice is supplied with regards to testing of the KB?

Are the testing facilities of the tool adequate to ensure proper validation of the application?

Where the KBS application is part of, or links with, a 'conventional' application/process, is guidance given on how such linkages may be established?

Does the KBS development methodology have to control development of traditional aspects or can they be developed in isolation and 'linked' at the end of development?

As methodologies, such as GEMINI, are only now emerging it is not surprising that we do not yet have tools to support the development process, other than the tools themselves which implement the knowledge. Investigation should be made here to determine if the supplier can provide aids for the analysis phase of KBS development, in particular the elicitation of knowledge

Chapter 5
Development methodology

from the expert. Inductive tools are, currently, the most common of these.

Are there any tools to support the process of eliciting the knowledge from the expert?

When using prototyping techniques, can the prototype be used as the basis of a system specification to be built using some alternative software?

5.2 Compatibility with GEMINI

It is important to determine that the supplier is aware of the status of GEMINI, and has made plans to ensure that the advice they offer clients with respect to the development of KBS applications will be compliant with that contained within GEMINI.

Has the supplier declared their intention to comply with the results provided by GEMINI?

Appraisal and Evaluation Library
Knowledge Based Systems Volume

6 Quality and control

This section is concerned with the facilities available for, and the degree to which it is possible to ensure, high quality software engineering. The development of a knowledge base system should be subject to similar controls in this area as traditional systems. To this extent the tool should provide facilities for ensuring that appropriate quality assurance measures can be applied to the development of the KBS. At present few KBS tools (as with Application Generator) offer any facilities in this area.

If a suitable approach is adopted to the development of KBS applications, such as that advocated by GEMINI, then conventional quality assurance techniques can be employed to ensure the quality of the KBS produced. With the stronger emphasis on the use of prototyping within the analysis and design stages the need for proper change control and version control is very important.

6.1 Documentation

Documentation should be 'active' rather than 'passive'. This is particularly important when controlled prototyping is taking place. The documentation for the prototype must be kept up to date.

What facilities does the tool have for rapid production of documentation (examples include the printing of object hierarchies and rule networks, and the provision of documentation facets for rules, objects and attributes)?

Can 'parts' of the KB (e.g. rules, objects, methods) be printed out, or is the facility restricted to the complete KB.

How is the printout structured?

Can this be tailored?

6.2 Development control

The KBS tool should support multiple versions of the application to exist concurrently; typically development, production and historic versions. Ideally this would be provided via a data dictionary (as with AG's).

Does the KBS tool allow the definition of development and production libraries?

Can historical versions of the KBS be maintained?

On-line interactive development can be expensive in terms of computing resources. Frequently, costing information relating to individual users/projects is required to provide a means of controlling access.

Does the KBS tool provide accounting information to enable the costing of development?

Version control and configuration management tools are increasingly in demand to control the development process.

Does the KBS tool provide mechanisms for managing 'versions' within 'releases' of the KBS application?

6.3 Audit control

The increased use of prototyping during the development of a KBS application makes it essential that there is recognition of 'production status', where access to knowledge bases is strictly controlled. In addition, when the knowledge contained within the knowledge base may represent a company 'asset', it may be desirable for audit and control purposes that the person seeking access be identified and the date and time be recorded.

It is likely to be an auditor's requirement that at least production libraries are protected against unauthorised access. Passwords are commonly used.

Can the production libraries be protected from unauthorised use?

The auditor may require to know when a program change has been effected. Ideally the system should write the date, time, user-id and program-id to a system log.

What information is logged when a knowledge base:

- *is accessed?*

- *is amended?*

6.4 Quality assurance monitoring

Because of the prototyping techniques employed within the analysis and design stages of developing a KBS application, it is important that the quality of the code produced is controlled and monitored. The minimum requirements are the ability to determine what prototypes exist, what their evaluation criteria are and who 'owns' them.

Does the KBS tool incorporate any facilities that can be used for Quality Assurance monitoring, for example utilities which scan or monitor knowledge bases and produce a statistical profile of where resources are used?

Can cross reference reports be produced of all knowledge base objects?

Appraisal and Evaluation Library
Knowledge Based Systems Volume

7 Portability

The ability to develop an application in one environment and deliver it in another is becoming increasingly important and applies to KBS as it does to traditional applications. At present there is no portability of code across tools, but there is portability of code from the same tool across different hardware platforms. The question then is how much effort is required to port code produced by the same tool between different machines on which the tool runs. Ideally the code would be 100% compatible, with a recompilation on the target machine being all that is required. An exception here relates to the products that provide toolkits for developing graphical user interfaces. If the code is developed on a workstation and ported to the mainframe environment then the user interface code will not be portable. This can be alleviated by providing character based emulators on the workstation version.

7.1 Environment independence

The hardware architectures and the operating system environments on which the tool runs should be determined.

On what hardware architectures does the KBS tool run?

On what operating system environment does the KBS tool run?

Between what hardware architectures and operating system environments does the vendor claim that code produced by the KBS tool is portable?

Are the same facilities/features provided across all environments?

7.2 Code portability

The availability of the KBS tool on multiple hardware architectures and operating system environments requires that code produced by the tool is portable across these environments with a minimum of effort.

How much effort is required to port code produced on one platform to a different platform, where the tool runs on both platforms?

If a graphical user interface toolkit is provided what is the process for developing both a graphical user interface and a character user interface to the same application?

Can character based user interfaces be ported without modification from one platform to another?

How is interaction with the environment (TP monitor or Operating System) on the target environment simulated on the development environment?

If databases to be accessed on the target environment do not run on the development environment how does this affect the portability of the code produced? How can the interface to the database be tested on the development environment?

7.3 Standards adherence

There is a consensus within the IT industry that the users of IT wish to move in the direction of an 'Open Systems Environment'. This is an environment in which the use of common interface definitions based on accepted standards facilitates the interconnection and interworking of different hardware and software products.

Does the vendor provide any commitment to conform to existing and forthcoming open systems' standards such as OSI, X/Open, Posix, SQL and SAA?

Where applicable, will the product adhere to emerging user interface standards, embodied by the 'look and feel' guidelines of, for example, Motif and Open Look?

8 End user interface

The user interface of the application is likely to be a major factor in achieving a successful implementation. While the 'look and feel' of the application is largely dependent on the skills of the development team, they are constrained by the functionality of the product. Depending on organisational or project requirements the tool should provide facilities for constructing either or both character based and graphical based user interfaces.

8.1 Character based user interfaces

In the mainframe environment it will be important for the tool to support rapid development of character based interfaces, through the use of a screen painter. The screen painter should support menus, forms, scrollable text windows and - if based on a workstation - hypertext.

Is there a screen painter for producing character based user interfaces?

Are menu, forms, and scrollable text windows provided for character based user interfaces?

Can the following attributes be specified for field attributes: protected, hidden (used for passwords), high brightness, reverse video, colour, blinking, font, character size, other (please specify)?

What types of menu selection are provided (ring menu, numeric designation, alphabetic designation, first character, scroll and highlight using cursor keys)?

Are function keys supported and programmer definable?

Can variants of the user interface be made available to different users or groups of users?
How can this be achieved?

8.2 Graphical user interfaces

The tool should take full advantage of the facilities offered by the windowing environment in which it operates. This entails providing full access to the graphical objects supported, such as radio buttons, checkboxes, menus, and full access to the window management system with the ability to create, delete, modify windows, and the use of a mouse.

A graphical object is used to display and/or retrieve attribute values or display information. It should be implemented as a system class, with the ability to create new graphical objects through inheritance. Typical attributes of a graphical object are: value, size, location, border, colour, and documentation string.

Are any graphical objects provided?

Can they be inherited to form new objects?

What attributes of graphical objects be accessed from rules and methods?

Is there a drawing facility for creating graphical based user interfaces?

What windowing environments does the tool support?

Does it support the full range of graphical objects supported by the environment, e.g. does it support buttons, radiobuttons, checkboxes, dialogue boxes, menus, creation/deletion/modification of windows, font changes, importing of graphic images?

8.3 Consultation facilities

Facilities for providing the user with both control over the consultation and access to the system's reasoning should be provided, e.g. HOW, WHY, WHAT-IF, HELP, SAVE/RESTORE.

Is an explanation facility provided?

Is a what-if facility provided?

Can help screens be programmed? Is browsing through the help system supported by the use of hypertext?

Can the help be made context sensitive?

What reporting facilities are provided?

Can a session be suspended, stored and restarted at a later date?

Chapter 8
End user interface

Do all users have to be given full access to the consultation facilities, or can they be restricted?

Appraisal and Evaluation Library
Knowledge Based Systems Volume

9 Security

It is important, due to the potentially confidential nature of the knowledge stored within a knowledge base that access to it can be controlled both during and after development. In particular an encryption facility should be available to ensure safe distribution.

The task of preventing users from gaining access to 'restricted' KBS should be handled by the operating system or TP monitor. If multi-level security is to be supplied the tools should provide a facility for communicating with the environment's security facilities via exit codes.

9.1 Access control

Are security issues concerning controlling user access, both to development environments and applications produced by the tool, handled by the operating system environment? If not how does the tool handle them?

How is multi-level security within an application handled?

Are there any additional user access issues involved in devloping an application using a range of software tools?

9.2 Encryption

Can the knowledge base be encrypted before going live on a multi-user system, or before being distributed?

Appraisal and Evaluation Library
Knowledge Based Systems Volume

10 Product credibility

Knowledge Based System tools are frequently the subject of 'imaginative' marketing. Some KBS tools are produced by small or relatively unknown software houses, or are written and supported in foreign countries and marketed here by agents, or represent initial forays into the market by well established software companies with a consequent lack of support. Other KBS are new to the market place and are therefore as yet untried. Such KBS are not necessarily of poor quality, but it is necessary to assess the likelihood of the software and the marketing agency still being maintained and in business respectively in the future before committing to using the product, irrespective of its technical merit.

10.1 Quality of product

At present this can only be established by talking with existing users and investigating if the developer subscribes to and complies with the requirements of BS 5750, which deals with a supplier's capabilities to operate a quality management system in the design, manufacture, installation, inspection and testing of a product.

What are the opinions of existing users?

Does the developer comply with ISO9001?

10.2 Product development status

Information on the development status of the product is essential before making a long term commitment to its use.

What is the current development stage of the product? For example:

- *static.*

- *stable but in the process of being cosmetically enhanced (ie minor changes and improvements in presentation, or in the way(s) in which the product interacts with the user, are in preparation - any such changes will not affect basic, user functions).*

- *being functionally enhanced.*

- *in the process of being developed for use on other machines.*

Have any enhancements been introduced recently? Are there any which are under development? If so, please list planned enhancements and the target dates for the introduction of such enhancements.

When was the last major, new version (as defined by the supplier) released and when is the next major, new version planned for release?

How does the company determine when a system requires enhancement and the nature of the additional or supplementary facilities which are to be incorporated?

Are there any known errors which have been identified in the current version of the product?

What plans are there for the product over the next 3 to 5 years? Will the product be different to today's version? If so, what differences will there be?

How is the system updating arranged to take account of new developments and legislation?

Have any overseas products been Anglicised (e.g. date format, £ symbol) or can they be altered by the user?

How many updates have been issued in the last year?

10.3 Supplier assessment

The supplier assessment will take into account the size of the supplier, whether they are the originators of the software or simply agents, how long they have been producing or marketing software, the size and whether they are a company based in the UK or abroad. The term supplier should be used in the widest sense, ie where the supplier is not the developer, all organisations involved in the development, marketing and support of the KBS should be investigated.

This section should help to identify KBS tools that are simply marketed by a supplier rather than developed, which has an obvious impact on the level of support provided.

Name, address, telephone number of supplier.

Chapter 10
Product credibility

Name(s), position(s), address(es) and telephone number(s) of the person(s) to contact for further details, if necessary regarding:

- *Marketing information*
- *Technical support*
- *Annual Report of company*

How long has the supplier been in operation:

- *in the UK?*
- *world-wide?*

Was the product originally developed by the above supplier?

What organisations have used the supplier's services in the past?

Does the supplier have a range of products covering related topics, ie is it an area in which they specialise?

Is the supplier a subsidiary of any other company? If so, please give details.

How many years has the supplier been active in the development and/or marketing of KBS tools?

During the last year, what percentage of total revenue has been derived from KBS tools?

What percentage of total profit or income has been contributed to research and development of KBS tools?

How many employees are dedicated to:

- *the development of KBS tools?*
  - *accounts?*
  - *sales?*
  - *other business ventures?*

What is the credibility of the MD and other key staff?

10.4   Product background   Most KBS tools start their life in a slightly unstable state. If a KBS tool has a reasonable number of production field sites (not simply out for approval or copies distributed but not used), then the product's capabilities and potential may be assumed to be at least adequate and the risk involved in selecting such a product is less than that of a new and untried package. New releases of a product may however be potentially risky.

Development Ancestry   Potential buyers should establish when the 'product' was first available rather than the concept.

When was the KBS tool first installed at a customer site for customer usage?

What is the source and history of product(s) under consideration?

For how long has the KBS tool been commercially available?

Did the supplier write the KBS tool, or are they acting as an agent?

Where is the software originator based, e.g. locally, UK, Europe, America?

Development profile   Many KBS tools are still in a state of development and enhancement. New features, facilities and environments are being added. While this may provide many useful new features it may cause problems if releases with desirable new features appear during development.

NB: New versions of KBS tools usually incorporate significant improvements is functionality or in performance. New versions are typically released on an 12 to 18 month cycle. Intermediate releases tend towards fixing bugs only.

How often are major product versions released? When was the last one released and when is the next one planned?

What enhancements, if any, are planned for the KBS tool and when will they be introduced? What is the supplier's policy towards compatibility between versions?

Chapter 10
Product credibility

How does the supplier determine when a system requires enhancement and the nature of the additional/supplementary facilities which are incorporated?

What techniques/methodology were employed in the development of the tool?

Product usage

An indication of the number of users of the KBS tool, the sales profile of the KBS tool and other pieces of information such as product appraisal and evaluation reports can give a valuable insight, as can visiting reference sites for the product.

How many user sites of the KBS are there:

- *in the UK?*
- *within Government?*
- *outside Government?*
- *elsewhere in Europe?*
- *elsewhere?*

How many systems of this type have been sold in the UK (and worldwide) during the last 12 months?

How many existing users are there (particularly any within government) and what is their volume of transactions (ie the number of applications currently in existence and use)?

For how long have earlier, original, users stayed with the KBS, or are all users (comparatively) recent?

Which is the nearest competing product available in the marketplace?

Describe any previous projects using this KBS tool with which the supplier has been involved both inside and outside of government. Please indicate the size and complexity of the jobs in broad terms.

Please give the name and address of reference site(s) which may be contacted if necessary.
Can other users be contacted, ideally in the same business area?

When was the first system of the type being considered (or proposed) successfully installed at a customer's site? Please provide details of the site's location and others, if available.

Does a user group exist for the product in question? If so please state:

- *whether it was formed independently of the supplier's organisation*
- *how long such a group has been in operation*
- *how active it is*
- *the number of active members*
- *joining/membership fees*
- *the number of meetings held each year*
- *when and where the meetings are held*
- *the name, address, and telephone number of the group's secretary*

How closely does the supplier collaborate with any such user groups which might be established?

Product information

Sources of information other than those suggested by the product supplier can be valuable.

Are there any independent reports and evaluations on the KBS tool being considered? Can copies of any reports be made available? If not, why not?

User profile

It is often claimed that users can build their own KBS applications with the high-level environments provided by KBS tools. In practice this tends to be restricted to very simple applications where the problem is of a procedural nature and the KBS tool provides access to its backward chaining inference engine without requiring an understanding of its other (if any) capabilities. Best results will always be attained when the correct tools have been chosen for the particular user profile. Some suppliers will claim usability by all classes of user.

Chapter 10
Product credibility

Who are expected to be the principal users of the KBS tool? For example:

- *non-IT staff*
- *analysts*
- *novice programmers*
- *experienced programmers*
- *AI Experts/Knowledge Engineers*

If non-IT users are expected to be able to use the tool ask for examples of this from existing users.

10.5 Documentation

KBS tools require adequate documentation. Frequently, at the beginning of the life of a KBS tool, the documentation is inadequate and/or poorly structured.

What information is available about the KBS tool before purchase?

What documentation is available, and how well is it presented?

What manuals and other documentation are provided when the KBS tool is purchased?

What other 'optional' manuals are available?

Can the documentation be copied by the user for their own use only?

What is the target audience for each manual?

Are the manuals available online? If so are they available in a hypertext format (on workstation versions)?

What information is available on the technical content of the system? For example:

- *database interfaces*
- *application program interfaces*
- *source code*

- *technique for indexing rules*

In the event of the supplier going out of business, what arrangements are there for access to the source code? For example, is a copy of the source code lodged with an Escrow Agent?

Do customers use the documentation provided or is there a need to develop instructions which are specific to each installation?

10.6 Training

The proper use of a KBS tool requires training, certainly for staff new to the use of the technology. For experienced users an introductory course often lacks the depth required. It is important that both Introductory and more advanced courses are available.

Please state whether training is included in the purchase price of the proposed software system and:

- *where such training is normally carried out*
- *whether onsite courses can be arranged*
- *the nature and amount of training normally required to operate and use the KBS tool (based on the supplier's previous experience)*
- *the duration of the training course(s)*
- *the individuals at which each course is aimed*

Are appropriate training courses provided by the supplier?

Do any third parties offer training in the use of the KBS tool?

How much computing expertise is required by the attendees?

Please describe any additional training related to the efficient and effective use of the system, such as one which covers design issues involving the use of the various paradigms supported by the tool, including details of cost, location, duration, and frequency.

10.7 Support

Support will be required, especially when the KBS tool is first introduced and before the organisation has developed its own inhouse expertise. The type and

Chapter 10
Product credibility

level of support available will depend on the size of the supplier organisation and the number of sites they are supporting. Support quality is also likely to be dependent upon where software development is carried out. If all development is performed overseas, then the local knowledge of the internals of the software may be reduced, with a consequent increase in the time taken to fix bugs.

General

Where and by whom is support undertaken?

Where are the support services located?

What is the policy for supporting previous releases of the KBS tool and how many versions are currently supported?

If users have access to the source code, to what extent is modification by users allowed without affecting support?

How are queries and problems dealt with after installation?

For which aspects of the implementation will the supplier be responsible?

Pre-sales

Is a demonstration available?

What are the arrangements for a trial of the KBS tool?

Does the right exist to reject the product if it fails user specified acceptance tests?

Will any verbal claims and promises made by the sales people be written into the standard contract?

Who will provide support/answer queries, and how accessible are they, e.g. by telephone, office hours only?

Installation

What maintenance and support services are available during installation of the KBS tool?

Will specific personnel be allocated to this project (full or part time; at the beginning of, during, and after implementation)?

Type and level

What maintenance and support services are available once

71

the KBS tool is operational?

How many technical support staff are supporting how many users?

How many technical support staff are available in the UK?

Does the supplier operate a 'hot-line' service for urgent enquiries and fault reporting? If so, is the service part of a system maintenance agreement and please state:

- *the average response time*

- *the longest response time*

- *the way in which the system operates.*

How long does it take for a supplier's hot-line to answer, and how long to resolve queries?

Is there a charge for the hot-line support?

What procedures are available for reporting problems and what action and priorities are assigned to rectifying faults?

Describe the circumstances in which onsite maintenance/assistance would be given. Would such services be provided by a sales representative or by a software expert/engineer? What would be the contractual response time for a call for assistance?

Fault correction

Are new versions of the KBS tool automatically sent to users?

What are the escalation procedures for fault correction? When will the Managing Director become aware of a serious fault?

10.8 Enhancements

The methods and procedures by which enhancements to software products are handled are extremely important in the context of reducing or avoiding disruption during the introduction of enhancements or improvements to the package.

What arrangements can be made for future changes which may be required by the user?

Chapter 10
Product credibility

What are the arrangements for future changes in requirements, and how will the work be costed?

How upward compatible is the KBS tool for changes to:

- *the hardware?*
- *the operating system?*

Are there facilities for users to 'customise' the KBS tool? If so, what are they?

Appraisal and Evaluation Library
Knowledge Based Systems Volume

11 Project specific criteria

These are any other requirements specific to the Departmental IT Strategy and/or project but not fully covered in other parts of this volume. In particular, this section could be employed to investigate the availability of domain or application specific KBS tools, AI programming languages or advanced Lisp-based tools.

Appraisal and Evaluation Library
Knowledge Based Systems Volume

12 Costs

It is usually a strategic objective to minimise costs but this is subject to meeting other requirements. Often costs are tangible but benefits are intangible. Within the CCTA User Guides it is recommended that cost should be calculated over the life of a study or project usually between 5 and 10 years. While cost comparison is performed in detail for the final selection of a product from the short list of approved products, there is also a case for including costing in the higher level formulation of the short list. For this purpose (ie software comparison), costing need not be done at a detailed or absolute level; approximate relative costs are sufficient.

12.1 Software

Software costs include a basic licence plus usually a recurrent annual maintenance charge. When prices are given it should be indicated whether these are inclusive or exclusive of VAT.

What is the basis of sale (ie lease, rental, purchase) and is the software the subject of copyright? Please provide details of licence fees which may be appropriate and the way(s) in which the licence operates, e.g. site licence, single system, total organisation? If a licence only covers the use of a package on a single, identified computer system, can it be run elsewhere in an emergency?

How much does it cost to buy the product outright? Does this cost include a copy of the source code?

Can the product be acquired on a trial basis and if so for how long and what are the costs involved? Are these costs discounted from any subsequent purchase price?

Does the product require any particular separately purchasable prerequisites or components (software or otherwise, from any vendor source)?

How much does it cost to rent or lease the product:

- *per month?*

- *per year?*

What are the minimum and maximum rental periods?

What are your terms for multiple copies of the product:

- *on a single site?*
- *on multiple sites?*

For the following customer support services, please give details of those which are available and the associated costs that are not included in the basic software charge:

- *design and configuration planning*
- *preliminary planning support*
- *development of clerical and operational procedures*
- *implementation*

If customisation of the package is available from the supplier, what is the basis for charges, and how many staff are available for such changes?

What happens if we buy your system but, after several months cannot achieve the benefits claimed? Has this happened to other purchasers?

What installation support is included within the purchase price? Does the price include the cost of new versions?

Is any warranty provided?

Is software support provided? If so what is the cost?

12.2 Hardware

If the product is to be installed on existing hardware, enhancement of this hardware may be necessary.

As with software, hardware is likely to have an initial capital cost together with recurrent maintenance cost. Costs may have to be considered for the system as a whole and not just for the package element.

What is the cost of enhancement of existing hardware (e.g. additional memory and/or disc space) in order to support the product?

Chapter 12
Costs

12.3 System development, operation and maintenance

Significant savings may be made by not purchasing unnecessary copies of development software, in an environment where applications are run on multiple sites, but development is done centrally.

Please specify the scope and terms of the maintenance contract; and describe those services which are standard and those which are optional.

What are the maintenance/support costs and arrangements for corrections, upgrades and new releases?

Does the maintenance charge cover the issue of new software releases/versions?

12.4 People

People costs are affected by factors such as the number needed, training requirements and their commercial worth. Sophisticated development tools reduce the number of people required for application development and probably reduces training costs. However, such trained people may be commercially attractive and require a high salary to retain them. Sophisticated tools may require skilled infrastructure support and such people are likely to be expensive. The uptake of new technology may require outside consultancy support that, while usually cost effective, will be expensive.

What is the cost of any training not provided free of charge when the product is purchased?

Please state whether training is included in the purchase price of the proposed system and:

- *where such training is normally carried out*

- *whether on-site course can be arranged*

- *the nature and amount of training normally required (based on your company's previous experience)*

- *the duration of training courses*

- *the individuals at which such training is aimed.*

Please describe any additional training related to the efficient and effective use of the system including details of cost; location; duration and frequency.

Is consultancy support available from the vendor? If so, what is the cost?

12.5 Documentation

Normally a package should be supplied with all relevant documentation necessary to operate it in an efficient and effective manner. Sadly some product's supplied documentation does not fulfil these requirements completely and additional documentation may need to be acquired. Some suppliers as part of their pricing policy may choose to charge separately for appropriate documentation.

What is the cost of any manuals not provided free of charge when the product is purchased?

Can manuals, once purchased, be copied for use only by the purchaser?

Criteria hierarchy

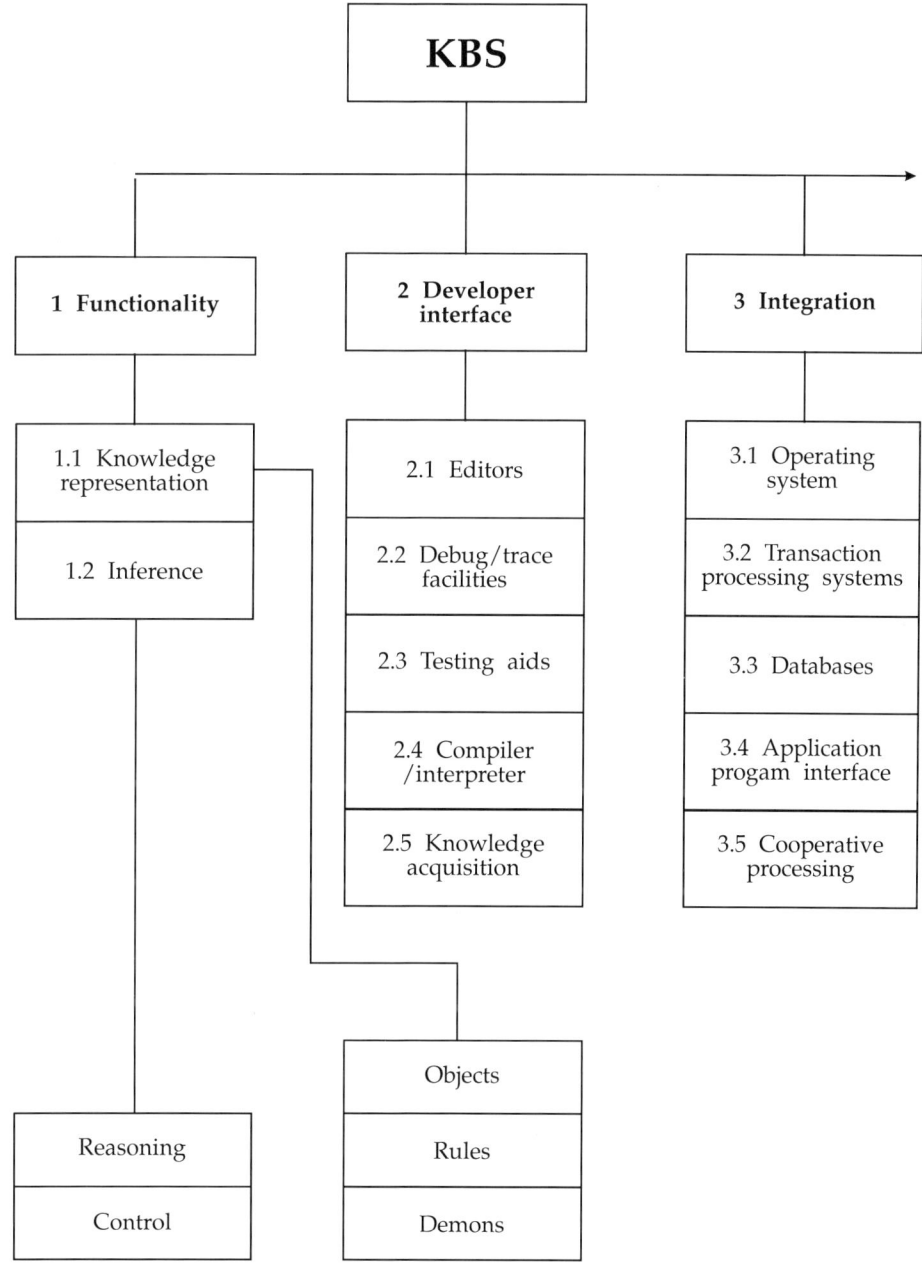

Appraisal and Evaluation Library
Knowledge Based Systems Volume

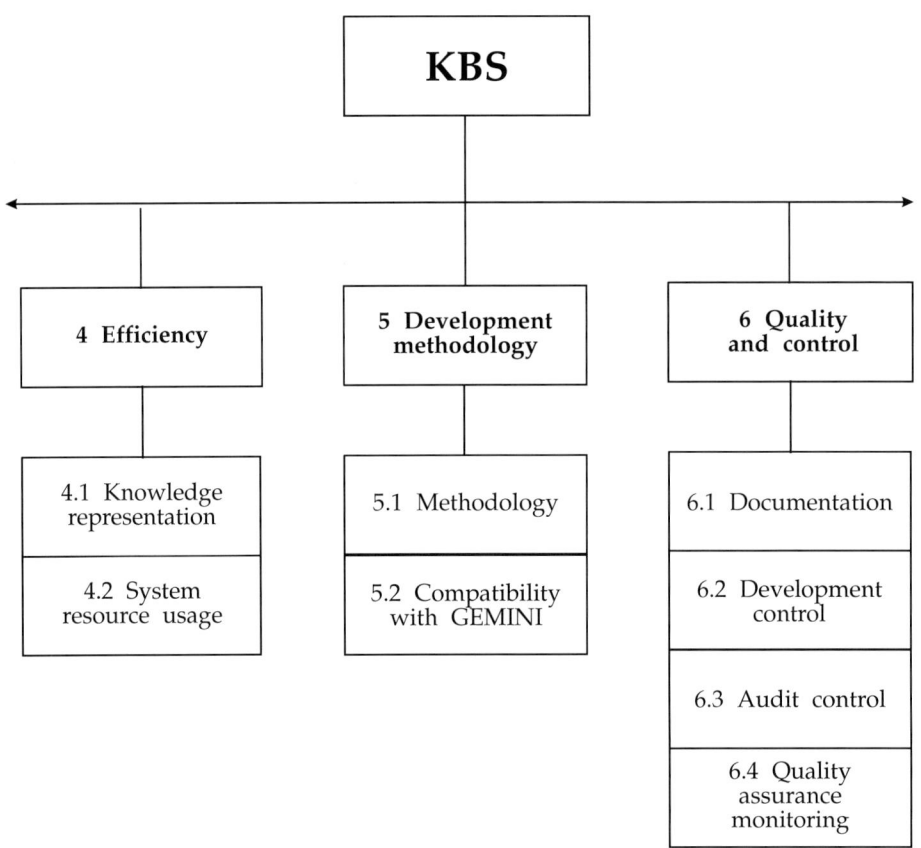

Annex A
Criteria hierarchy

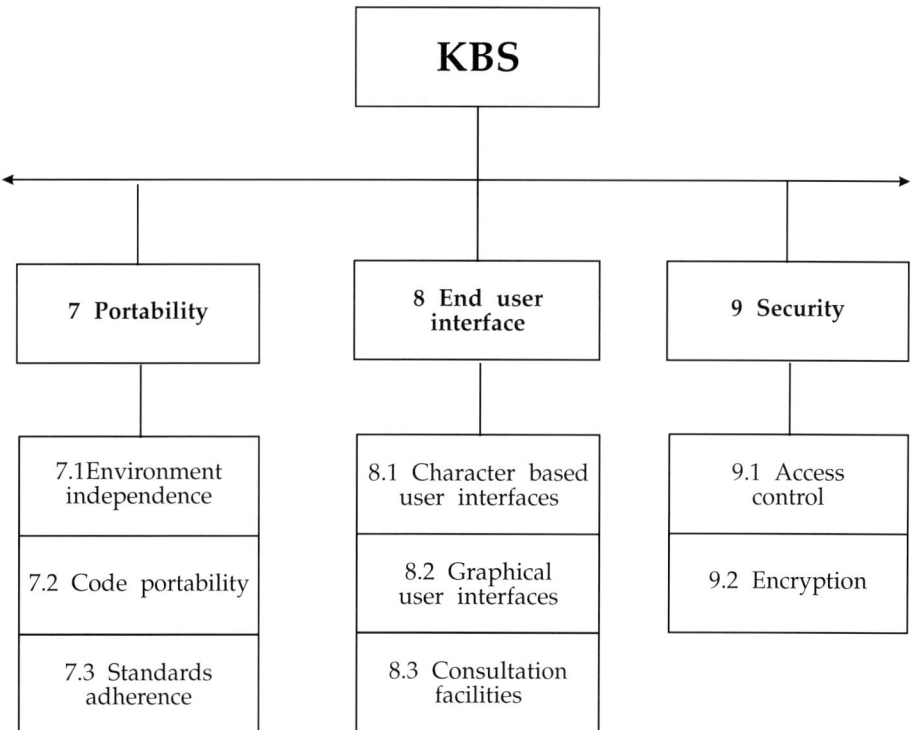

83

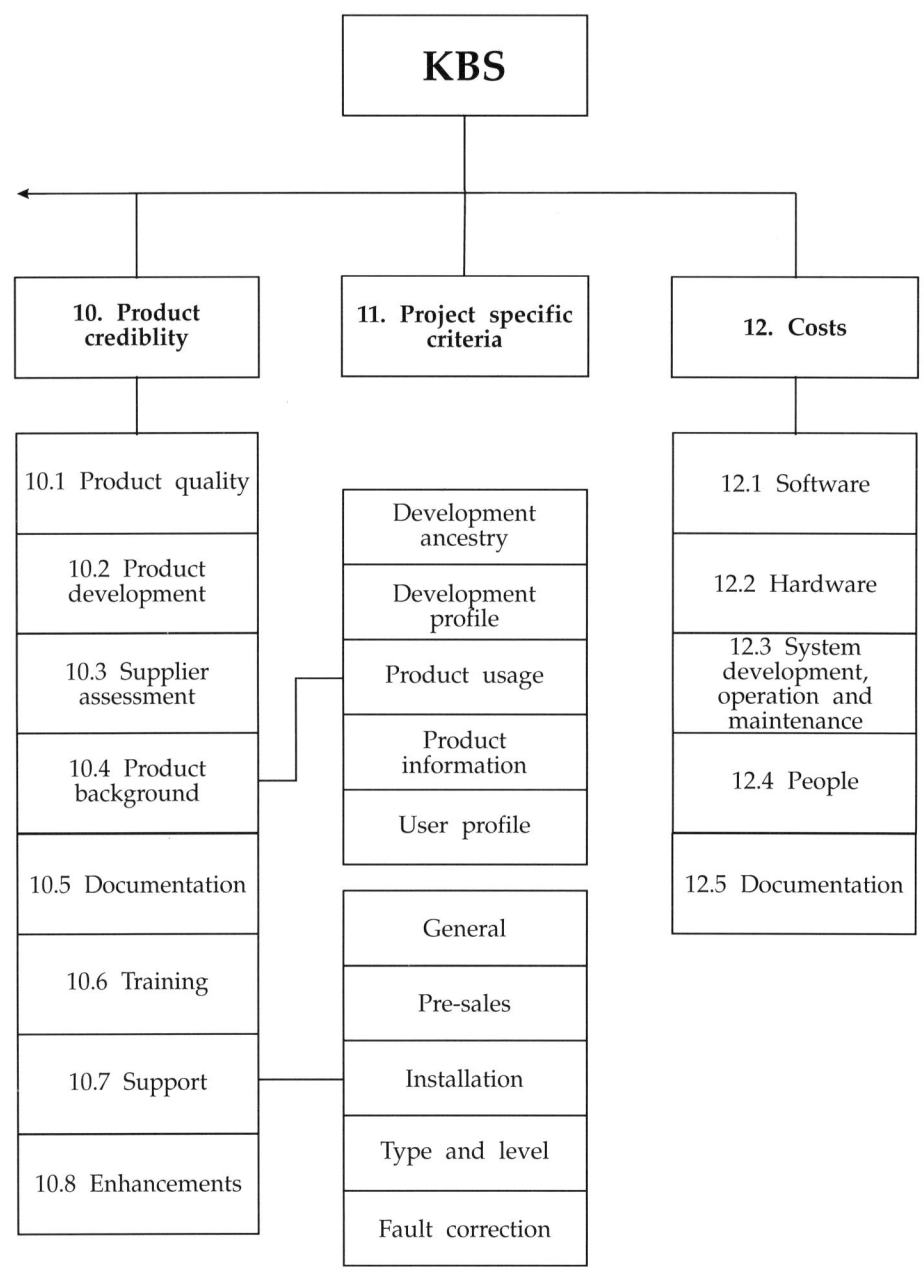

References/additional reading

Artificial Intelligence (2nd edition),
Patrick Henry Winston,
Addison-Wesley,
1984.

Artificial Intelligence and the Design of Expert Systems,
George F. Luger and William A. Stubblefield,
Benjamin/Cummings,
1989.

Developing Expert Systems,
Edmund C. Payne and Robert C. McArthur,
John Wiley and Sons,
1990.

Encyclopedia of Artificial Intelligence,
Stuart C. Shapiro (ed),
John Wiley and Sons,
1987.

Programming Expert Systems in OPS5,
Lee Brownston and Robert Farrell and Elaine Kant and Nancy Martin,
Addison-Wesley,
1985.

Printed in the United Kingdom for HMSO.
Dd. 295353, C5, 9/91, 3390/3, 5673, 162098.

Further information

Further information on the contents of this module can be obtained from:

Environmental Infrastructure Services
CCTA
Riverwalk House
157-161 Millbank
London
SW1P 4RT
Tel. 071 217 3182 (GTN 217 3182)

Further information on the contents of the IT services management modules of the IT Infrastucture Library can be obtained from:

IT Infrastructure Management Services
CCTA
Gildengate House
Upper Green Lane
Norwich NR3 1DW
Tel. 0603 694854 (GTN 3014 4854)

Further information on the services available from Property Holdings Fire Branch can be obtained from:

Fire Policy and Standards Branch
Directorate of Special and Central Services
DOE Property Holdings
Apollo House
36, Wellesley Road
Croydon CR9 3RR
Tel. 081 760 8757 (GTN 3831 8757)

The price of this publication has been set to make some contribution to the preparation costs incurred by the department.

Printed in the United Kingdom for HMSO
Dd292699 9/91 C7 G3390 10170

IT Infrastructure Library
Fire Precautions in IT Installations

OTHER COMMENTS

Return to: Environmental Infrastructure Services
CCTA
Riverwalk House
157 - 161 Millbank
LONDON SW1P 4RT

IT Infrastructure Library
Fire Precautions in IT Installations

Comments sheet

CCTA hopes that you find this book both useful and interesting. We will welcome your comments and suggestions for improving it.
Please use this form or a photocopy, and continue on a further sheet if needed.

From:
            Name

      Organization

          Address

         Telephone

COVERAGE
Does the material cover your needs?
If not, then what additional material would you like included.

CLARITY
Are there any points which are unclear?
If yes, please detail where and why.

ACCURACY
Please give details of any inaccuracies found.

If more space is required for these or other comments, please continue overleaf.

Annex E
Halon - an effective, clean and safe fire extinguishing agent?

The development strategy sought to meet all of the following criteria of acceptability:

* retention of critical technical and performance factors associated with existing products, particularly agent cleanliness

* ozone depletion potential of less than 0.05, preferably zero

* toxicologically acceptable for use as a fire fighting agent

* accessible manufacturing route at acceptable cost on appropriate timescale.

In ICI's opinion, it is unlikely that an alternative for halon will be found in the near future. The likelihood is that the process will take ten years or more to achieve the requirements for a cost-effective, efficient, safe and low - or zero - ozone depleting replacement for halon.

A major programme of work is under way in the USA to find a replacement for halon. The United States armed forces are evaluating a range of new chemicals. Although some have proved to be very effective fire extinguishants, they would contribute to the greenhouse effect.

Du Pont, the major manufacturer of halons in the USA, is to introduce two new fire extinguishants - a hydrofluorocarbon called HFC-125 and a hydrochlorofluorocarbon called HCFC-123. Both these chemicals are considered potent greenhouse gases. HCFC-123 will also contribute to the depletion of the ozone layer, although its ODP has been calculated at 0.02 compared with 10 for Halon 1301.

Figure E.2: Ozone depletion potential of CFCs and halons

	Compound	Ozone depletion potential
Group 1 compounds	CFC 11	1.0
	CFC 12	1.0
	CFC 113	0.8
	CFC 114	1.0
	CFC 115	0.6
Group 2 compounds	Halon 1211	3.0
	Halon 1301	10.0
	Halon 2402	6.0

E.4.4 Ozone Depletion Potential (ODP)

The Montreal Protocol established a list of ozone depleting substances and rated them according to the capacity to reduce the ozone in the upper atmosphere. Extracts from this list appear in Figure E.2.

The Protocol allows for the production or consumption of a substance within Group 1 or Group 2 to be switched so long as the overall ozone depletion potential is not increased.

E.5 Possible long term alternatives to halons

Although halons do not pose the same threat to the environment as the CFCs (very much smaller quantities are released), the long term future of halon as an extinguishing gas is in doubt. ICI Chemicals and Polymers Limited, amongst others, are actively seeking alternative fire fighting agents.

A paper presented by ICI Chemicals and Polymers Limited at the Firex North '89 Conference titled **A future for halons? - Possible long-term alternatives** outlined the strategy adopted by the company to develop such alternatives.

E.4.2 Montreal Protocol

A substantial breakthrough in helping to protect the environment was the signing in 1987 of the Montreal Protocol on Substances that Deplete the Ozone Layer. The Protocol, which came into force in January 1989, sets limits on both the production and consumption of CFCs and halons.

For CFCs this involved freezing consumption at the 1986 levels by mid 1989; a reduction of 20 per cent by 1993-94; and a further reduction to achieve an overall 50 per cent cut-back in consumption by 1998-99.

For halons no progressive cut-backs were proposed but simply a freeze in consumption at 1986 levels by 1992.

In 1990 the Montreal Protocol was amended to reduce the consumption of halons to 50 per cent of 1986 levels by 1995 and, except for essential applications where no alternatives are available, to phase out halons completely by the year 2000. The Protocol will be reviewed again in 1992 and further restrictions are possible.

E.4.3 CFCs and halons

CFCs belong to a family of compounds known as halogenated hydrocarbons or halocarbons. They are hydrocarbons in which some or all of the hydrogen atoms have been replaced by atoms of one or more of the group of elements known as the halogens: fluorine, chlorine, bromine and iodine. CFCs are halocarbons made up of chlorine and fluorine.

The halons are also halocarbons but comprise compounds with bromine as well as fluorine and chlorine. Bromine acts like chlorine in the destruction of ozone but, on a per atom basis, is even more destructive and thus halon is implicated in the depletion of the stratospheric ozone.

CFCs are used as aerosol propellants, foam blowing agents, solvents and refrigerants. The principle use of halon is as a fire fighting agent.

An American report (Halons and the Stratospheric Ozone Issue) comparing the production of CFCs 11, 12 and 113 and Halon 1211 and 1301, indicated that the halons only accounted for 1 per cent of the total. The report also estimated that only 30 per cent of Halon 1301 and 15 per cent of Halon 1211 manufactured in 1985 was actually released into the atmosphere.

Halon 1301 is accepted for use in protecting manned areas provided that the installation is designed in accordance with **BS 5306: Part 5.1. BS 5306: Part 5.1** recommends a maximum concentration of 6 per cent for Halon 1301 and 4 per cent for Halon 1211 (because of its higher vapour toxicity) for use in manned areas. Halon 1211 is not generally used for total flooding systems to protect normally occupied areas.

E.3.2 Discharge hazards

Hazards associated with discharge include:

* noise - lasting for 10 seconds, this is startling rather than harmful

* turbulence - loose papers and lightweight objects can be scattered and unsecured ceiling tiles can be dislodged

* cold temperature - direct contact with the discharging gas could cause frostbite.

E.3.3 Decomposition by-products

Halon 1301 decomposes upon contact with fire and hydrogen halides are produced. The concentration of these gases, produced as a by-product of the extinguishing process, is very small in relation to other potential sources of hydrogen halide gases in computer accommodation. For example: PVC insulation and CFCs used in air-conditioning units.

From whatever source, hydrogen halide gases are corrosive and the protected area should be ventilated as soon as possible to protect equipment from contamination and corrosion.

E.4 Ozone depletion potential of halons

E.4.1 Background

In the 1970s scientists first propounded the theory that certain man-made products known as chlorofluorocarbons (CFCs) were having an effect on the earth's ozone layer. The CFCs were being broken down by sunlight and the chlorine released was destroying ozone in the stratosphere. In 1985 an Antarctic survey team discovered a hole in the ozone layer which provided the first real evidence to support the theory.

Annex E. Halon - an effective, clean and safe fire extinguishing agent?

E.1 Introduction

A great deal has been written about the use of halon as an extinguishant and this annex addresses some of the topical issues, particularly in relation to the use of halon in computer installations. Is the concern about toxicity and the corrosive properties of halon justified? What effect has the 1987 UN Agreement (Montreal Protocol) to reduce CFCs by 50 per cent had on the halon market?

E.2 History

Halons 1301 and 1211 were originally investigated by the United States Army following the second world war in a search for more effective, less toxic fire extinguishing agents than, for example, carbon tetrachloride. Halon 1301 was found to be one of the most effective agents evaluated and had the lowest toxicity of them all. Halon 1301 was selected for further development and since that time its use has expanded into a wide range of military, commercial and industrial applications. In the 1950s a study by United States Air Force indicated that Halon 1211 had good properties for hand held extinguishers. Promoted (and manufactured) by ICI, the use of Halon 1211 spread rapidly within Europe, the applications being similar to those for Halon 1301.

E.3 How hazardous is Halon 1301?

E.3.1 Toxicity

Halon has been tested extensively for toxicity. In high concentrations and for prolonged exposures it can produce temporary dizziness, irregular heartbeat and eventually unconsciousness and death. As a precaution, recommended exposure times (by the Health and Safety Executive) are low and these are detailed in Figure E.1.

Figure E.1: Halon 1301 - recommended maximum exposure times

Concentration	Max. exposure time
7% or lower	15 minutes
7-10%	1 minute
10-15%	30 seconds
greater than 15%	exposure should be prevented

D.3 Reference detector

An additional reference detector can be used to take into account the effects of external pollution. The effects of external pollution can be subtracted so that only a net gain from internal sources will produce an alarm.

D.4 The HART integrated monitoring system

The HART integrated monitoring system combines high sensitivity air sampling smoke detectors, individual cabinet air sampling units, point ionization, heat and optical detectors with manual alarm call points and audible warning devices, all operated by a common system controller.

Automatic and manual actions can follow a pre-programmed fire plan relating to all alarm, evacuate, shutdown and extinguishing procedures stipulated by the client, the insurance company or other professional advisor.

Annex D. The HART detector

D.1 Introduction

This annex includes a technical description of the Hartnell HART detector.

The HART detector has been designed as part of an integrated monitoring system to give advanced warning of fire in computer rooms and telecommunications buildings. It detects the presence of very small concentrations of smoke presented to it by air sampling pipe work and a small fan.

The basic concept is very similar to the VESDA, but the equipment incorporates more recent technology.

D.2 Principle of operation

Air samples are drawn along a pipe network or from existing air supply ducts to the detector. Particles in the sample are illuminated by a miniature diode laser and the scattered light is detected by a single photon avalanche detector. The flow of air through the instrument is measured and the resulting signal is fed, together with the signal for particle (or smoke) concentration, to the control equipment. A signal processor produces a smoke intensity signal which is displayed on the control unit.

Sensitivity

The equipment is capable of detecting diluted smoke at concentrations as low as 0.01 per cent obscuration per metre.

Staged alarms

Three independently adjustable alarm thresholds may each be set to operate a staged warning system as follows:

1. ALERT - may simply alert a responsible person

2. ACTION - could activate an evacuation system or shut down overheating equipment

3. FIRE - would operate alarms and may call the fire brigade and release an extinguishing medium.

To prevent short term variations in the environment causing false alarms, each staged warning output has an adjustable delay timer. Only a sustained increase in smoke level will result in an alarm.

When smoke is detected the following operations occur:

* the activation of audio and visual alarms

* the closing of relay contacts to facilitate automatic power cut-off to affected areas (cabinets or rooms)

* if applicable, the discharging of a fire extinguishant into protected cabinets (through the smoke detection pipe work) or the execution of total flooding of the protected space; where total flooding is installed, alarms from two separate zones must be received before the discharge is activated after a preset time delay.

C.4 Installation

The STAMP system is particularly suitable for computer suites. Difficult points in such accommodation, such as equipment cabinets, floor and ceiling voids and air handling units are easily protected using a STAMP system. The sampling heads are easy to install and no electrical connections are required in the floor or ceiling.

Annex C. The STAMP fire detection system

C.1 Introduction

This annex includes a technical description of the Single Tube Automatic Multi-Point (STAMP) fire detection and extinguishing system. Originally developed to meet the needs of the Ministry of Defence (Navy) and the Fire Research Station, STAMP provides an aspirating system where air samples are drawn through small bore pipes from the areas being monitored to the detector.

C.2 The detector

The detector is based on enhanced ionization chamber technology. In ionization detectors a number of processes take place concurrently in the same volume, ie:

* air is ionized by a radioactive source
* the ionized air reacts with any smoke particles present
* any remaining ions are extracted by an electric field, the resulting current giving a measure of the density of the smoke.

Normally in such detectors there is only a short period of interaction between the ions and the smoke so that small quantities are difficult to detect. The STAMP detector is more advanced. Air is drawn through a long tubular chamber with the ionizing source at one end and the detector at the other. A delay is introduced between the initial ionization and the measurement, lengthening the period of reaction between the smoke and the ionized air with a resulting increase in detector sensitivity.

C.3 Principle of operation

Sampling is carried out using a network of small-bore tubes (typically 6 mm polyethylene) connected through a manifold assembly to the smoke detector and a vacuum pump. An arrangement of solenoid valves opening and closing according to a preset programme allows up to twelve zones to be sampled sequentially, the sequencing cycle being continually repeated automatically. A flow check unit monitors the effectiveness of the vacuum pump and the sampling tubes.

Annex B
The VESDA System

Coincidence detection   **BS 6266** recommends the provision of 'coincidence' detection before activating the release mechanism of an automatic extinguishing system. The VESDA can be used very effectively in conjunction with point detectors to provide the required independent confirmation. The VESDA can operate as a pre-alarm device, to bring to the attention of the user that a change has occurred in the computer environment. Trained personnel should be able to act on this information, investigate the cause of the alarm and take any necessary action.

2. The intermediate threshold, FIRE ALERT, could activate a personnel evacuation system or disconnect overheating electrical equipment.

3. The highest threshold, FIRE ALARM, would operate the fire alarm sounders, call the fire brigade and/or release the extinguishing medium (if combined with automatic fire extinction).

The alarm thresholds are programmed on-site to smoke levels higher than usual.

Where external pollution may occasionally enter the building, upsetting the internal air quality, an additional detector (for reference purposes) may be installed at the inlet air point. This enables the effect of external pollution to be subtracted so that only a net gain in smoke from internal sources is detected.

In certain situations brief periods of increased smoke levels may be quite usual without necessarily indicating a fire. To prevent such 'false' alarms, independent delay timers are provided for each alarm threshold. Only a sustained increase in the background smoke level beyond the time delay will result in an alarm output. Another source of false alarms for conventional smoke detectors are air drafts associated with air-conditioning. Drafts are not a problem with the VESDA; continuous air movement is a requirement. An inbuilt flow meter provides an 'airflow failure' alarm should airflow fall below levels required for rapid response.

B.3 Installation design considerations

The manufacturer of the VESDA has drawn attention to certain aspects that should be considered when designing an automatic smoke detection system incorporating the sensitive aspirating detectors.

Point detectors

The conventional point smoke detector has been in existence for more than 40 years and has reached a high level of stability and reliability. The performance of the detector in still or low velocity air is extremely good. However, in computer areas where the airflow can be high, experience has shown that smoke can be pulled away from detectors by the air handling equipment. An automatic fire detection system, however, must also work effectively with the air conditioning turned off. Under these circumstances the point smoke detector will provide a high degree of protection.

Annex B. The VESDA system

B.1 Introduction

This annex includes a technical description of the Very Early Smoke Detection Apparatus (VESDA) system.

VESDA is an optical airborne particle monitoring instrument, configured for early warning fire detection. The instrument is several hundred times more sensitive than conventional smoke detectors. This acute sensitivity can detect incipient fires before smoke concentrations become dangerous and before property damage is significant.

B.2 Principle of operation

Samples of the atmosphere from the protected area are continuously drawn along a sampling pipe network to the detector by means of an in-built fan. Particles within this air sample are exposed to an intense light source within a precision optical chamber. Photons scattered off the molecules of air and particles of smoke are captured by a solid state receiver. The resultant signal is amplified and processed to provide an analogue output of airborne smoke intensity. The VESDA controller module provides a display of this smoke intensity level.

Sensitivity

In most optical systems a beam of light is transmitted across an area in which smoke may be generated. Particles of smoke would obscure some of the light causing a lower light intensity at the receiver. Obscuration caused by incipient fires is typically 0.1 per cent per metre - that is a 0.1 per cent change in received light intensity compared with clean air over a distance of 1 metre.

The VESDA is capable of detecting smoke concentrations as low as 0.01 per cent obscuration per metre (0.1 per cent obscuration/metre full scale deflection). For use in comparatively smoky environments detectors are available with less sensitivity such as 0.2, 0.5, or 1.0 per cent full scale deflection.

Alarm levels

There are three independently programmable alarm thresholds, each of which may be set to any level on the display unit. This enables the operation of a staged warning system, for example:

1. The lowest threshold, STAFF ALERT, may simply alert a responsible person.

The IT Infrastructure Library
Fire Precautions in IT Installations

Definitions used in the module

AFD or AFX system	A fire detection or extinguishing system that, under specified conditions, functions without the intervention of a human operator.
coincidence detection	A facility designed in such a way that an output is obtained only when at least two independent detection inputs are present at the same time. (Also known as 'double knock'.)
Common User Estate	Mainly comprises office accommodation in large towns or cities, managed by DOE Property Holdings on behalf government departments occupying the accommodation.
fire door	A door, for people to use, which is intended, when closed, to resist the passage of fire and/or gaseous products of combustion.
fire extinguishing medium	The substance contained in a fire extinguisher or fire-extinguishing system that, when discharged onto the fire, will put it out.
fire resistance	The ability of a component or construction of a building to satisfy, for a stated period of time, the appropriate criteria specified in the relevant part of **BS 476 Fire tests on building materials and structures**.
fire stop	A non-combustible seal provided to ensure a close fit between elements in a building (for example, a cable passing through a floor) so as to restrict penetration of smoke or flame.
halons	A family of fire extinguishing gases.
manual fire alarm system	An electrical system in which the alarm is initiated manually from one of a series of manual call points with break-glass covers. The actual warning is produced by the electrical operation of suitable sounders.
non-combustible	Any material (that does not burst into flames) capable of satisfying the performance requirements specified in **BS 476: Part 4 Non-combustibility test**.
soffit	The underside of beams, enclosed staircases etc.
sprinkler system	An assembly of pipework, graded in size, erected throughout a building, and on which sprinklers are mounted at prescribed intervals. The pipework is connected to a set of system control valves incorporating a hydraulic alarm and is fed by an approved water supply.
total flooding system	A system in which a gaseous extinguishing medium is discharged into an enclosure to a concentration sufficient to ensure extinction of a fire.

Annex A. Glossary of terms

Acronyms and abbreviations used in the module

AFD	Automatic fire detection
AFX	Automatic fire extinguishing
BCF	Bromochlorodifluoromethane (Halon 1211)
BS	British Standard
CEN	European Committee for Standardization
CFC	Chlorofluorocarbon
CIBSE	Chartered Institute of Building Services Engineers
CRAMM	CCTA Risk Analysis and Management Methodology
DOE	Department of the Environment
EC	European Council
EN	European Standard (prefix used by CEN)
FIRTO	Fire Insurers' Research and Testing Organization
FOC	Fire Officers' Committee
FPA	Fire Protection Association
GOCO	Government owned company operated
HSE	Health and Safety Executive
ICI	Imperial Chemical Industries
IEE	Institution of Electrical Engineers
IT	Information Technology
LPC	Loss Prevention Council
NFPA	National Fire Protection Association (USA)
ODP	Ozone depletion potential
PABX	Private automatic branch (telephone) exchange
PH	Property Holdings
PSA	Property Services Agency
PVC	Polyvinylchloride
STAMP	Single Tube Automatic Multi-Point
UN	United Nations
VDU	Visual display unit
VESDA	Very early smoke detection apparatus

PSA Performance Specification PF2.PS - Platform Floors. (PSA 1990)

Regulations for Electrical Installations. Fifteenth Edition. (Institution of Electrical Engineers 1981)

13. Bibliography

BS 5306 - Code of practice for fire extinguishing installations and equipment on premises. (British Standards Institution 1986)

BS 5266 Part 1 - British Standard code of practice for the emergency lighting of premises. (British Standards Institution 1988)

BS 5445 Specification for components of automatic fire detection systems. Part 7 - Point type smoke detectors. (British Standards Institution 1984)

BS 5588 - Fire precautions in the design and construction of buildings. Part 3 - Code of practice for office buildings. (British Standards Institution 1983)

BS 5839 - Fire detection and alarm systems for buildings. Part 1 - Code of practice for system design, installation and servicing. (British Standards Institution 1988)

BS 6266 - British Standard code of practice for fire protection for electronic data processing installations. (British Standards Institution 1982)

CRAMM (CCTA Risk Analysis and Management Methodology). (CCTA 1988)

Draft British Standard code of practice for fire protection for electronic data processing installations (Revision BS 6266: 1982). (British Standards Institution 1990)

Fire detection and alarm systems - A guide to the BS Code BS 5839: Part 1. P Burry. (Paramount Publishing Limited 1988 - ISBN 0 947665 11 0)

Fire precautions in computer installations. M B Wood. (NCC Publications 1986 ISBN 0 85012 556 1)

FIREX Midlands '88 Conference Papers. (Paramount Exhibitions and Conferences 1988)

FIREX North '89 Conference Papers. (Paramount Exhibitions and Conferences 1989)

1st International FIREX 1990 Conference Papers. (Paramount Exhibitions and Conferences 1990)

IT in the Civil Service No.4 - Accommodation for computer and office systems. (CCTA 1983. ISBN 0 11 630722 6)

Automatic fire detection and extinguishing systems offer no guarantee of successful protection in the event of a fire. Systems can fail to work, due to, for example, a component failure, or a loss of gas concentration. It is essential that the design, installation and maintenance of AFD and automatic fire detection (AFX) systems is undertaken by experts in this field.

Halon 1301 is the most effective fire extinguishing agent for computer rooms, but the environmental acceptability of its continued use is very much in question. Although halons do not pose the same threat to the environment as CFCs (most of the halon manufactured is banked and only released in cases of emergency) it is possible that the manufacture and use of halons will be banned by the turn of the century. As a result, heavy costs could be incurred for modifying halon flooding systems in the future.

Section 12
Benefits, costs and possible problems

12. Benefits, costs and possible problems

12.1 Benefits

The main benefits of following the advice in this module will be:

* reduced probability of a fire, because of well defined fire prevention procedures

* confidence that the IT installation is equipped with the level of fire protection equipment commensurate with the risk to the installation and the business

* improved integrity of an organization's IT services; the main benefit of active fire protection (for example, a smoke detector system), is the time that it buys - even short exposure to smoke from PVC can cause irreparable corrosive damage to IT equipment

* knowledge of the appropriate actions to be taken in the event of a fire.

12.2 Costs

The widespread use of extensive, and expensive, fire protection systems in IT installations arises from the consequences of loss due to fire and not from a high probability of fire or a significant threat to life.

After a fire, the consequential losses (interruption to business, loss of profit, wasted staff time and so on) can be many times greater than the material losses incurred in the fire. Such potential losses should be taken into account when determining the most suitable fire protection for a computer centre or office.

At a cost, there is a solution to almost every fire protection problem.

12.3 Possible problems

Perceived problems of following the guidance in the module relate primarily to automatic fire detection and extinguishing systems.

Faulty equipment, high levels of airborne dust or infiltration of smoke from a fire in an adjoining building can lead to a high number of false alarms from automatic fire detection (AFD) equipment. People will stop responding if there are frequent false alarms.

11.3 Fire Research Station

The Fire Research Station forms part of the Building Research Establishment, an Executive Agency of the Department of the Environment. The basic remit of the Station is to provide advice, supported by research, to government on all aspects of fires and fire protection.

11.4 Loss Prevention Council

The Loss Prevention Council (LPC) is the UK organization involved with all aspects of fire and loss prevention, and loss control worldwide. The Technical Centre of the Loss Prevention Council combines the resources of the former Fire Officers' Committee (FOC) and the Fire Insurers' Research and Testing Organization (FIRTO). The LPC's Technical Centre provides a wide range of facilities for testing, evaluation, approval and certification of fire safety equipment, products and materials used in the protection of buildings.

11.5 Fire Protection Association

The Fire Protection Association (FPA) is supported by insurance companies and is recognized by government, fire service and industry as a national independent fire safety information and advisory organization. Government departments and major industrial and commercial associations are represented on the FPA Council, together with insurance interests. The FPA is a constituent company of the Loss Prevention Council.

11. Support services

Support services (including professional advisory services, research into all aspects of fire, testing of fire protection equipment) are available from a wide variety of organizations. These include the following.

11.1 DOE Property Holdings and PSA Services

On 1 April 1990, the Property Services Agency (PSA) was divided into two separate departments - DOE Property Holdings and PSA Services. On 1 April 1991, PSA Services became a government owned, company operated (GOCO) organization preparatory to becoming a private concern.

DOE Property Holdings was set up to act as the landlord for government buildings and land holdings. While managing the common user part of the civil estate, a significant proportion of that estate (some 40 per cent) is now the responsibility of individual occupying departments.

Within Property Holdings the Fire Branch offers a professional advisory service to government organizations on fire protection matters.

PSA Services has two divisions from which government organizations can commission services: PSA Projects deals with project management and design for major projects and PSA Building Management handles the smaller works schemes, maintenance and estate surveying services. However, government organizations are completely untied from PSA Services and are free to test the market for those services.

11.2 Health and Safety Executive

The Health and Safety Executive (HSE) was created by the Health and Safety at Work etc Act 1974 to inspect places of work and enforce legislation. The Health and Safety Commission, established under the same Act, is responsible for developing legislation and for general policy on health and safety. As an example, the HSE has issued a guidance note (GS 16) covering the safety of gaseous fire extinguishing systems.

The IT Infrastructure Library
Fire Precautions in IT Installations

	* any automatic fire alarm and extinguishing system should be tested and serviced in accordance with manufacturers' recommendations and **BS 5839: Part 1** and **BS 5306**.
Fire fighting equipment	Portable fire extinguishers should be maintained in working order and checked and tested in accordance with **BS 5306: Part 3**. This action can be undertaken either in-house or by the fire appliance companies.
Housekeeping	Waste paper and other refuse should not be allowed to accumulate. It should be stored in a separate location and be removed from the building daily.

Particular attention should be paid to keeping paper and other combustible materials away from equipment that is left running unattended (often after the end of the working day).

Vital records (paper, magnetic and optical) should be safeguarded against loss in a fire by storage in cabinets or safes. Copies of essential records should also be kept in another building.

Security staff should be briefed about their responsibilities for fire precautions and the action to be taken upon discovering a fire both during and after office hours.

10.5.3 Training

All staff should be trained to understand the fire precautions and the action to be taken in the event of a fire. Training generally should provide for:

* the action to be taken on discovering a fire
* raising the alarm
* the action to be taken on hearing the fire alarm
* the correct method of calling the fire brigade
* the location and use of fire-fighting equipment
* knowledge of escape routes
* appreciation of the importance of keeping fire doors closed
* stopping machines and isolating power supplies where appropriate
* evacuation of the building.

A practice fire drill should be carried out at least once a year.

Fire Instruction notices should be displayed in prominent positions throughout the building. The notices should state, in concise terms, the action to be taken on hearing the fire alarm and on discovering a fire.

10.5.4 Routine fire precautions

The following general precautions should be undertaken:

* the premises should be inspected daily to ensure that all escape routes are clear and that all doors are unlocked or are capable of being opened in the event of a fire
* all parts of the premises should be checked regularly for fire hazards and to ensure that fire doors are closed
* a final inspection should be undertaken at the end of the working day and all non-essential electrical equipment should be turned off at the mains

Halon 1211 (BCF) is a very efficient agent for use on fires involving live electrical equipment. However, it should be noted that halons and their decomposition fumes are likely to be hazardous to people in enclosed spaces with restricted ventilation, for example, floor or ceiling voids, and small congested service equipment rooms. Halon extinguishers present little danger when operated in the open air, large rooms or other well ventilated areas.

Although less efficient than BCF extinguishers, carbon dioxide extinguishers are still very effective on fires in electrical equipment and carbon dioxide is more acceptable environmentally than the halons. Dry powder extinguishers are not recommended in office areas.

Siting of portable extinguishers

Portable extinguishers should be placed in clearly visible and accessible positions. It should be unnecessary to have to go more than 30 metres from the site of a fire to the nearest extinguisher. Extinguishers should be sited in similar positions on each floor.

The increase of IT in the office environment has resulted in a significant increase in the amount of electrical equipment located in office buildings. However this equipment does not present a special fire risk and there is no requirement to site additional extinguishers near IT equipment.

10.5 Operational fire precautions

10.5.1 General

Management and staff should be fully aware of their duties and responsibilities for fire prevention and the action to be taken in the case of a fire. Detailed guidance on this subject is given in **BS 5588 Part: 3 Appendix A - Advice to management and procedure in the case of fire**. Guidance from that publication is summarized in the following paragraphs.

10.5.2 Responsibilities

A senior official (with necessary deputies) should be appointed as the fire precautions officer, with responsibility and authority for ensuring that:

* fire precautions are satisfactorily maintained
* adequate instructions are issued to staff
* training and fire drills are undertaken.

Section 10
Fire precautions in the office

10.4 Fire protection facilities

Mandatory requirements for the provision of fire protection (fire warning systems, fire fighting equipment, etc) are contained in the **Fire Precautions Act 1971**.

10.4.1 Fire warning

Fire alarm systems

The most common fire warning system, installed in all but the smallest office building, is the manual fire alarm system. This is an electrical system in which the alarm is initiated manually from one of a series of manual call points with break-glass covers. The actual warning is produced by the electrical operation of suitable sounders.

Fire detection and alarm systems

In some cases, a fire alarm system alone may not meet the compulsory requirements for fire safety. It will not cover the protection of property in buildings left unattended at night time and in parts of buildings only visited occasionally. In such cases a fire detection and alarm system could greatly reduce the time between the outbreak of a fire and its discovery. Fire alarm or fire detection and alarm systems should be designed in accordance with **BS 5839: Part 1**.

10.4.2 Fire extinguishing systems

Automatic fire extinguishing systems

The value of installing automatic fire extinguishing systems in offices is to minimize any delay in starting fire fighting. Water sprinkler systems have been used for many years and there is really no alternative with equal advantages. A sprinkler system will ensure that the spread of the fire is controlled, prevent danger and summon the fire brigade automatically. The disadvantages of water sprinklers are covered in Section 8.5.3.

The design of sprinkler systems should be in accordance with **BS 5306: Part 2**.

Sprinkler systems are not generally installed in government offices.

Manual fire fighting equipment

There is a legal requirement (**Fire Precautions Act 1971**) to provide fire fighting equipment for use by the occupants of the building. Hose reels and portable fire extinguishers serve for this purpose.

Fire extinguishers suitable for dealing with fire in electrical equipment should be the only type sited close to such machines. Carbon dioxide and halon extinguishers can both be used safely on electrical fires.

The majority of computer terminals were linked to remote processors, separately protected in specialized accommodation. Where a fire was contained locally the loss was probably not great.

10.3.3 The office of the 1990s

Equipment installed in the 1990s office is not significantly different. There are still photocopiers, word processors, computer terminals and facsimile machines. The trend is towards intelligent workstations (based on personal computers), probably networked with each other and using common data storage.

There are more potential sources of electrical ignition in offices but, provided that there is a satisfactory standard of basic management, electrical maintenance and housekeeping, the risk of fire is not now significantly greater.

However, organizations are now far more dependent on IT and the importance of apparently minor installations can be seriously underestimated. The real change in risk comes in the nature of the consequences of a fire. Organizations should investigate how serious a fire would be in particular circumstances. For example: data communications between remote offices will probably depend on the telephone system - and essentially on the PABX, which itself may be a computer. If the exchange catches fire then the business of the organization could be seriously affected.

Current causes of fires are much the same as they have been in the past, that is, failures to recognize potential fire risks and responsibilities.

As we have seen, the risk from incipient electrical equipment fires remains low. However, the introduction of a large number and variety of electrical and electronic items into the office has tended to result in a degree of familiarity which can breed carelessness. Office systems have not eliminated paper from the office. Paper handling needs to be disciplined. For example, paper should be stored away from equipment and waste promptly removed.

Smoking continues to start fires and provision should be made for the safe disposal of smoking materials.

Section 10
Fire precautions in the office

10. Fire precautions in the office

10.1 Introduction

VDUs, printers and other equipment installed in office accommodation fall outside the scope of **BS 6266**. The fire precautions that should be adopted in office buildings are covered in this section.

Most office accommodation is outside the control of IT divisions and the need to ensure that disciplined contingency plans are in place, to facilitate recovery of office IT systems, is addressed in this section.

10.2 BS 5588

Fire precautions in the design and construction of buildings Part 3: Code of Practice for office buildings.

This code of practice is cited in the **Building Regulations 1985**. The code provides detailed guidance for designers and building construction teams on measures that should be taken in the event of a fire to safeguard the lives of employees and of the public in office buildings. These measures may help protect the building and its contents in the event of a fire.

10.3 The fire risk

10.3.1 Case histories

All large fires are logged at the Fire Protection Authority (FPA). The number of recorded fires that started in offices rose from 7485 during the period 1974 to 1980 (an average of 1247 a year) to 8559 during the period 1981 to 1986 (an average of 1426 a year).

During the period 1982 to 1986 there were 31 large office fires where individual losses amounted to more than £250,000. The total material loss exceeded £24M. Consequential losses were not recorded.

10.3.2 The office of the 1970s and early 1980s

Offices generally used conventional furniture and office machinery included typewriters, word processors, photocopiers, calculators and probably computer terminals. Only rarely would the machinery itself start a fire, and when it did, the characteristics were those of an electrical fire.

The IT Infrastructure Library
Fire Precautions in IT Installations

Section 9
Operational fire precautions

9.7 Contingency measures

9.7.1 Back-up copies

Copies of important records (systems and application programs and data files) should be kept in a secure location, which would not be affected by the same fire incident. Copies should be taken in a controlled way at regular intervals so that an effective recovery could easily be achieved.

9.7.2 Contingency plan

Organizations should draw up a contingency plan that includes detailed procedures to ensure that appropriate actions can be taken to enable full recovery from back-up storage media. The contingency plan should be a controlled document and a copy should be kept with the back-up copies of the important records.

The IT Infrastructure Library module on **Contingency Planning** provides a comprehensive guide to producing a contingency plan and recovering from a disaster.

* when and how to operate any fire extinguishing system

* when and how to operate emergency or master switches to isolate power to computer equipment and air-conditioning.

9.5 End-of-the-day fire precaution routines

Fires do occur in unoccupied computer suites either where staff have left the premises at the end of the day or where equipment is operated unattended. Clear instructions should be given to designated staff on actions to be taken before leaving the premises. These instructions should include the following:

* check that all doors and hatches between rooms are closed

* remove all waste paper and other combustible materials

* unplug from sockets all ancillary and miscellaneous equipment not required in unoccupied hours

* put any automatic fire detection and extinguishing system on automatic control

* return all appropriate records to store rooms and safes.

9.6 Automatic fire extinguishing systems

Automatic gas extinguishing systems may put the health and safety of staff at risk. If correctly designed, installed, maintained and operated, the risk to staff is low. However, it is recommended that staff are evacuated and doors closed before the gas is discharged.

For systems using carbon dioxide, Halon 1211, or Halon 1301 in concentrations above 6 per cent by volume, no person should be allowed to enter the room until the automatic discharge feature is locked off. **BS 5306** gives detailed guidance on safety precautions relating to carbon dioxide and halon extinguishing systems.

9. Operational fire precautions

9.1 Introduction

Although computer installations have a low inherent fire risk, the consequences of a fire are so serious that there is a need for high standards of fire protection. Prevention of fire is important since even a small fire, which may be extinguished quickly by a gas flooding system, can result in a significant financial loss. Good operational fire precautions reduce the likelihood of a fire. This section of the module outlines procedures to be followed.

9.2 Smoking

Smoking should be prohibited in the computer suite and 'NO SMOKING' signs should prominently be displayed. Ash trays or sand bins should be made available immediately outside prohibited areas. Provision should be made for staff rest areas, including smoking, outside the air-conditioned area.

9.3 Staff training

Staff should be trained in the fire precautions for the site, the procedures for dealing with fires and the use of hand-held fire extinguishers. It is important to test procedures and a test should be undertaken at least once a year. Halon should not be discharged during training exercises.

9.4 Fires during working hours : procedures to be followed

Clear instructions on the action to be taken when a fire occurs (reminding staff of what they have been told and have practised in training) should be displayed in the computer area. These instructions should include:

* how to raise the alarm
* procedures for calling the fire brigade
* evacuation procedures
* how to notify key personnel
* how to use hand-held extinguishers, fire blankets etc, without taking personal risks

expensive leak detector. In addition it contributes to the depletion of the ozone layer and non-essential emissions should be avoided.

Procedures are now available that provide information better than that obtained from discharge tests. An example is the enclosure integrity test, details of which are outlined in Section 8.6.2.

Substitute test gas

Where it remains necessary to conduct a discharge test, then a substitute test gas should be used. A number of substitute gases exhibit properties similar to halon. A test programme undertaken in USA has shown that sulphur hexafluoride (SF_6) is an excellent simulant for Halon 1301. This gas is commercially available in the UK.

8.6.2 Enclosure integrity test

The enclosure integrity test is a procedure developed to locate the source of leaks and predict retention time without discharging the halon. The test is conducted with a device known as a door fan. It was developed by the energy conservation industry to determine the 'leakiness' of buildings and thus their ability to retain heat. There are three basic components:

* a panel system to fit in the doorway
* a calibrated and variable-speed fan
* pressure gauges.

The test is conducted by blowing air into the room to develop the same pressure difference that would be created by the weight of the halon. By measuring the air flow required it becomes possible to measure the equivalent leakage area (equal to the total of the cracks, gaps and holes in the room). The equivalent leakage area can be related to the room volume allowing correlation with the results of actual Halon 1301 discharge tests.

This procedure has the advantage of permitting enclosure integrity tests to be undertaken before the system is installed. Leaks, which can be detected easily with chemical smoke, can then be sealed before the halon system is commissioned.

Further information on halon room integrity testing is available from the UK distributors of the Canadian manufactured test equipment (see details in Section 13).

Section 8
Putting out fires

Halon 1301 and water sprinklers

Computer cabinets can be protected by the direct injection of halon into the cabinets. The floor void can be protected by a total flooding halon system, while the structure of the building and the ceiling void can be protected by a water sprinkler system. A significant saving of halon can result from such a combination of extinguishing methods.

Carbon dioxide and water sprinklers

To eliminate the use of halon, a total flooding carbon dioxide system can be used to protect the floor void, and a water sprinkler system to protect the structure, floor void and contents within the enclosure. However, the response of water sprinkler systems is slower than gaseous systems and as a result increased damage may be incurred. Consequential water damage may also be significant.

8.6 Acceptance testing of gaseous total flooding systems

BS 5306: Section 5.1 covers the tests that should be undertaken on completed Halon 1301 total flooding systems. These tests include a full discharge test (using Halon 1301).

8.6.1 Discharge test

A discharge test has been used in the past to prove or disprove that a system will operate and meet its design criteria. A test implies pass/fail criteria as follows:

* discharge duration - the discharge to be substantially completed in 10 seconds (or as short as reasonably possible to ensure that the extinguishing concentration is reached rapidly to minimize the effect of halon breakdown)

* time to reach uniform concentration - within 60 seconds of the end of discharge (the speed of diffusion can be greatly affected by the location of the discharge nozzles)

* hold time for the concentration - 10 minutes (related to the time taken for the fire brigade to arrive on site).

The main cause of discharge test failures is the inability of the protected space to hold the necessary concentration of gas for the required time. This is nearly always because the protected space is not sealed properly. Halon is an

BS 5306: Part 2 also deals with:

* water supplies, their pressure and flow rate
* basic principles of design of the system
* pumps, tanks and other components
* materials, workmanship, inspection, testing, approval, maintenance and repair procedures.

Disadvantages

The disadvantages of sprinkler systems are:

* the need to provide an adequate water supply; systems now being installed may require higher water pressure and flow rates than the mains can provide and this will often mean installing pumps and tanks
* the response of water sprinkler systems is slower than gaseous systems
* water does not have the permeability of halon or carbon dioxide
* water can do considerable damage if applied for any length of time.

8.5.4 Foam and powder systems

Foam systems

Foam systems have been used to protect large numbers of cables in cable tunnels and, occasionally, in floor voids in computer installations. Foam systems suffer many of the disadvantages of water sprinkler systems and, in addition, chemical dispersants are required to break down the foam during the cleaning up after a discharge.

Powder systems

Powder systems should not be used in computer installations (see Section 8.2.6).

8.5.5 Combination of systems

Halon 1301 is the most effective fire extinguishing gas for use in computer installations. However the environmental acceptability of halon as a fire suppressant in the long-term is in doubt. By combining a number of alternative systems it is possible to provide a level of protection comparable with halon (without endangering life). Such systems will help to reduce or eliminate the use of halon as a fire suppressant in computer installations.

Section 8
Putting out fires

* evacuation alarms should be easily distinguishable from normal fire alarms

* adequate escape routes and exits should be provided, clearly identified and kept clear at all times

* a means of ventilating the protected area, after the discharge of gas, should be provided

* people should not be permitted to enter the protected area after a fire or a gas discharge until it has been declared safe by a responsible authority.

Storage space

The high liquid density of halon, coupled with its high extinguishing effectiveness, results in a significant reduction in storage space requirements for systems using this agent compared to those using carbon dioxide. A Halon 1301 system requires less than one-fifth of the storage volume of an equivalent carbon dioxide system.

8.5.3 Water sprinkler systems

To avoid water damage to equipment and loss of IT operations whilst drying out, most computer manufacturers recommend keeping water sprinkler systems away from computer suites.

In some instances insurance companies require the building structure to be protected by automatic sprinkler systems. Such systems may be specified as a long stop in case a gas flooding system fails to extinguish a fire.

A sprinkler system consists essentially of an array of closed sprinklers mounted on pipework under the ceiling of the protected space. Each sprinkler head contains an inbuilt thermal detection unit and, when activated, will discharge a spray of water onto the fire below. Sprinkler response time depends on the temperature rating of the fusible element in the detection unit.

BS 5306: Part 2 specifies several types of system including:

* wet; the pipes are permanently charged with water

* pre-action; the pipes are normally dry and fill with water when a fire operates a separate detection system; for water to be discharged it is necessary for the sprinkler heads to operate in the normal manner.

The pre-action sprinkler system minimizes the possibility of accidental discharge of water due to failure or damage to the automatic sprinkler heads.

and any equipment fire can be extinguished in its earliest stages. In addition to the considerable reduction in the amount of gas required, there are equipment and installation cost savings. The equipment supplier should be consulted before fitting in-cabinet protection.

Benefits

The benefits of gas extinguishing systems are:

* fires are extinguished very rapidly (provided that a reliable fire detection system is installed to actuate the system), reducing losses and cleaning up from the actual fire to a minimum

* the gas will permeate throughout the protected space and will reach even difficult locations

* carbon dioxide and halon systems, if correctly designed, installed and used by trained staff, present a low hazard risk to people

* the testing of the entire system can be carried out easily (by use of room pressurization tests), at moderate cost and without any consequential cleaning up costs

* reliability and performance of gaseous extinguishing systems are good and continue to improve as a result of the implementation of comprehensive codes of practice

* accidental discharges do not result in prolonged downtime.

Safety

The following steps should be taken to increase the safety of the system:

* audible and visual alarms should be provided in the protected area to indicate, where applicable:

    - operation of the fire detection system (audible and visual alarm)

    - start of time delay period (audible alarm)

    - commencement of extinguishing agent discharge.

* systems should be equipped with a MANUAL ONLY/ AUTOMATIC AND MANUAL selector switch; for systems using carbon dioxide or Halon 1211, no person should be permitted to enter the protected area until the automatic feature of the fire extinguishing system is locked off

Section 8
Putting out fires

8.5 Design considerations for automatic fire extinguishing systems

8.5.1 General

The design of AFX systems for computer installations is covered in detail in **BS 6266**. An AFX system is operated by a suitable method of fire detection (see Section 7.4). The computer installation should be provided with one or more of the following:

* a gaseous flooding system using halon or carbon dioxide
* the ability to apply either halon or carbon dioxide directly into equipment cabinets
* an automatic water sprinkler system.

8.5.2 Gas extinguishing systems

Gas extinguishing systems (the installation and principal components) should conform to the recommendations in the following British Standards:

* **BS 5306: Part 4 Carbon dioxide systems**
* **BS 5306: Section 5.1 Halon 1301 total flooding systems**
* **BS 5306: Section 5.2 Halon 1211 total flooding systems.**

Halon 1211 total flooding systems are not generally installed in computer accommodation.

Total flooding systems

The design of total flooding systems should ensure that a uniform concentration of gas is achieved throughout the space to be protected. Care must be taken that the required concentration of gas can be maintained for long enough to extinguish and prevent re-ignition of the fire. All doors should be close fitting and provided with gaskets all round. Any space around cable and service entries to the protected area should be sealed and low-level ventilation ducts should be fitted with effective closures.

In-cabinet protection

An alternative to total flooding systems is to inject the gas directly into the equipment cabinets. Installing a smoke detector in the equipment cabinet ensures a rapid response

8.3 Portable fire fighting equipment

Portable fire extinguishers of the carbon dioxide or halon type (in accordance with **BS 5423 Specification for portable fire extinguishers**) should be provided in conspicuous and accessible positions throughout the computer installation.

Halon 1211 (bromochlorodifluoromethane - BCF) is the gas most widely used in hand portable extinguishers for use on fires involving electrical equipment. Carbon dioxide, although less efficient, is an effective alternative to halon and is less damaging to the environment. Halon and carbon dioxide extinguishers present few problems when operated in well ventilated areas but should not be discharged in small enclosed areas. A 2.5 kg halon or carbon dioxide extinguisher discharged in a small office would result in an overall concentration of less than 5 per cent.

Dry powder extinguishers can be used for electrical fires.

Water extinguishers are more effective than halon or carbon dioxide extinguishers on carbonaceous fires, that is, those involving paper, plastic or other carbons. Water extinguishers should not be located in computer rooms, but should be provided near paper stores and adjacent to equipment handling large quantities of paper.

Staff working in the computer suite should be trained in the safe use of available fire fighting equipment.

8.4 The need for automatic fire extinguishing systems

The need for automatic fire extinguishing equipment depends on the following factors:

* operational staff levels - if the site is operated unattended (at any time) then a gaseous flooding system may be justified

* the time for competent action to begin once a fire alarm is raised, including how long it takes the local fire brigade to get there

* the risk to the installation - organizations should establish the risk and the level of measures to be taken, for example, by using **CRAMM**.

Section 8
Putting out fires

* is electrically non-conductive
* has excellent penetration
* causes minimal direct and secondary damage.

There are some disadvantages with halon as a fire extinguishing agent:

* halon decomposes on contact with fire and the by-products may be more hazardous than the un-decomposed gas
* halon is one of the gases contributing to the depletion of the ozone layer.

Volatility   The high volatility of Halon 1301 is advantageous in total flooding systems, where a rapid and uniform distribution of the gas throughout the effected space is important. A large proportion of Halon 1301 is discharged initially as a gas. Any remaining liquid vaporizes very rapidly on exposure to the warm air in the effected space. The lower volatility of Halon 1211 makes it very suitable for use in hand held portable extinguishers from which the agent is applied directly to the fire. Since a large quantity of Halon 1211 is discharged as a liquid, it can be directed effectively at the fire.

The use of halon in automatic fire suppression systems for computer accommodation is considered in more detail in Section 8.5. The future use of halon is reviewed in Annex E.

8.2.5 Foam

Foam consists of a bubble-structure, formed by aerating and agitating a solution of concentrate in water, and works by blanketing or smothering a fire. Most foam solutions can cause corrosion and their use is restricted largely to the extinguishing of flammable liquid fires.

8.2.6 Powder

Powders are finely divided chemicals (for example bicarbonates of sodium or potassium) which are used to extinguish fires in flammable liquids, gases and solids. Powders discharged onto equipment combine with atmospheric moisture and become mildly corrosive. In addition powder melts on hot surfaces and then hardens when cool, leaving a cleaning up problem.

Concentrations of between 5 and 10 per cent can cause breathing difficulties and a concentration of 10 per cent is regarded as the danger level for most people.

When released, carbon dioxide discharges as a cold vapour which has an initial cooling effect. Adding a minimum of about 30 per cent of carbon dioxide to the air feeding the fire reduces the oxygen content and smothers the fire.

Carbon dioxide is a very effective fire extinguishing agent. The main advantages are that it:

* is fast acting
* is electrically non-conductive
* has no effect on materials and no cleaning-up problems
* permeates throughout the protected space when used in a total flooding system.

The use of carbon dioxide in automatic fire suppression systems for computer accommodation is considered in more detail in Section 8.5.

8.2.4 Halon

The halons (often referred to simply as halon) are colourless, odourless gases. Some sixty halons have been tested and all show some degree of fire suppressing ability. Halons extinguish fires by a chemical reaction that inhibits combustion. Two halons (1211 and 1301) are currently used as commercial-grade fire extinguishing agents.
A concentration of as little as 3 per cent of Halon 1211 or 1301 is effective against burning solid fuels and 4.5 per cent will deal with most flammable liquids.

Safety

Current knowledge (Health and Safety Executive document **GS16**) suggests that exposure to approximately 4 per cent concentration of Halon 1211 and 6 per cent concentration of Halon 1301 for several minutes gives no ill effect to people. Higher concentrations or longer exposures are potentially hazardous.

Halon is a very effective fire extinguishing agent. The advantages are that it:

* is non-toxic at extinguishing concentrations for short periods
* disperses readily and therefore causes no safety problems

8. Putting out fires

8.1 Introduction

The types, effectiveness, problems and dangers of the various fire extinguishing substances (water, gas, foam and powder), together with their use in portable and fixed extinguishing systems, are reviewed in this section.

This section also looks at factors that need to be considered when assessing the need for automatic fire extinguishing systems and reviews the implications of installing such systems.

8.2 Substances used for putting out fires

8.2.1 General

Extinguishing substances available are:

* water

* gaseous extinguishing media, including carbon dioxide and the halons (halogenated hydrocarbons)

* foam

* powder.

8.2.2 Water

Water is cheap, widely available and the most used extinguishing medium. Water cools the burning surface and inhibits the development of the fire. Water is non-toxic and is very effective for extinguishing Class 'A' fires (fire involving solid material usually of an organic nature). However water has the following significant disadvantages:

* it conducts electricity and should not be applied to live electrical equipment

* it can cause considerable secondary damage if used for any length of time.

8.2.3 Carbon dioxide

Carbon dioxide is a colourless, odourless gas which is present in the atmosphere at a concentration of 0.03 per cent by volume. A minimum localized concentration of 30 per cent is generally required to extinguish a fire.

The IT Infrastructure Library
Fire Precautions in IT Installations

Section 7
Detecting fires

7.4.4 Operation of AFD system

The AFD system should:

* operate alarms (audio and visual) in the computer room and in other areas where action should be taken

* switch off all power to the room except for the lights and smoke extract equipment

* shut down the air conditioning and close dampers in ductwork

* initiate release of extinguishing substance (where installed), subject to a pre-determined time delay to enable people to leave the protected area before the discharge begins.

Loss of power or automatic cutouts can cause considerable disruption to IT operations. To reduce the possibility of false alarms triggering automatic cutouts, the power-down sequence can be delayed until two detectors (on separate circuits) are activated.

7.5 Acceptance testing

Testing of the completed fire detection system should be undertaken to verify the correct siting and operation of the smoke detectors and the operation of the alarm system. Tests should be carried out with the air-conditioning system on and off to ensure that the response of the smoke detectors meets all requirements. Details of procedures and simulated fires are defined in **BS 6266**.

detail in **BS 6266**. A mixture of equal numbers of ionization and optical point detectors should be used. Smoke detectors should comply with the requirements of **BS 5445 : Part 7**. The spacing of detectors in peripheral offices, stores, corridors etc within the installation should comply with **BS 5839: Part 1**.

Optical beam smoke detectors can be used to monitor room space and should be used in conjunction with ionization chamber smoke detectors.

Even with a concentration of detectors, air flow patterns in a computer room can affect the operation of the detector system. Care should be taken to position point detectors to ensure fast detection rates.

Aspirating smoke detector systems provide very high sensitivity and can detect incipient fires. This type of detector is gaining acceptance for use in protecting computer suites but owing to the various methods of application available it has not been practicable to set standards for sensitivity and arrangements. A British Standard for these systems is in preparation and should be available towards the latter half of 1991. However the high sensitivity of these systems can lead to false alarms and manufacturer's design and installation recommendations need to be taken into account.

Three examples of systems based on the aspirating detector are:

* the VESDA detector, distributed by Fire Fighting Enterprises
* the STAMP fire detection system, manufactured by Guardian Fire Detection Systems Limited
* the HART detector, manufactured by Kidde Hartnell

Brief descriptions of these systems have been included in Annexes B,C and D.

7.4.3 Detector zoning

Detectors should be zoned to facilitate the identification of the origin of a fire, for example, under floor and ceiling voids. If detectors are in hidden areas, a diagram showing their position should be placed in a suitable location (for example, near the operators' console).

Section 7
Detecting fires

temperature or when the temperature rises at an abnormally rapid rate. They are unsuitable as primary detectors because their response is substantially slower than smoke detectors. Heat detectors can be used to actuate fire dampers.

7.3.4 Flame detectors

Flame detectors respond to the radiation (infra-red and ultra-violet) from flames. They have a very rapid response time and provide a useful addition to smoke detectors, but are usually only installed for special hazards such as those in chemical plant or flammable liquid storage.

7.4 Design considerations for automatic fire detection systems

7.4.1 General

The design of fire detection systems for computer installations is covered in **BS 6266**, which in turn references **BS 5839**.

The types and positioning of fire detectors are important factors to be considered to ensure that fires are detected early and false alarms avoided.

It is essential that the design, installation and testing of AFD systems is undertaken by professionals.

7.4.2 Detector type and spacing

When determining the type, density and positioning of detectors in computer rooms the following factors have to be taken into account:

* the room will contain expensive equipment that could be extensively damaged by even a small fire; detectors must be positioned to ensure that any fire is detected as soon as possible

* air-conditioning creates turbulence and can dilute smoke from a fire before it reaches the detectors; the detectors must also be effective with the air-conditioning switched off.

Recommendations for the density and positioning of point-type smoke detectors in computer accommodation (which includes space within false floors and ceilings, air-conditioning systems and ventilation ducts) are covered in

Ionization chamber detectors	This detector consists of an ionization chamber containing two electrodes, across which a potential difference is maintained, and a radioactive source. The radioactive source, by ionizing the air, produces positive and negative gas ions which travel to the electrode of opposite polarity, resulting in a current flow. If smoke particles enter the chamber, the charged ions attach themselves to some of the particles. As the charged smoke particles are heavier than the gas ions they move more slowly between the electrodes. As a result the charged particles take longer to meet particles or ions of the opposite polarity and thus are neutralized. This leads to a reduction in the current flowing in the chamber and sets off the alarm.
	Ionization chamber detectors are most sensitive to smoke containing small particles - clean burning fires with little smoke.
Optical point detectors	Optical detectors rely on the obscuring and/or scattering of light by smoke particles. In an optical point detector a light beam is focused on or near a photo-electric cell. When smoke is present the output from the photo cell changes and this sets off the alarm.
	Optical point detectors respond well to smouldering fires.
Optical beam detectors	This detector uses a beam of light that will detect fire at any point along its length. It consists of a pulsed light source producing a focused beam which is detected (perhaps after reflection) by a photo cell some distance away. When the beam is attenuated by smoke or irregularly deflected by air convection above a fire, an alarm is given.
	Optical beam detectors perform well when there is rapid dispersal of smoke.
Aspirating detectors	With this type of detector air is continually drawn through the device and the concentration of smoke is measured. The current range of aspirating detection systems is more sensitive than other smoke detectors, resulting in shorter response times. Such systems are particularly effective in the early detection of smoke. Since electrical fires typically begin with smoke, aspirating detectors have an obvious application in computer suites.

7.3.3 Heat detectors

Heat detectors operate using a variety of physical principles - including the expansion of metals, liquids or gases, changes in electrical characteristics and the melting of solids. Heat detectors respond either at a fixed, preset

7. Detecting fires

7.1 Introduction

Effective fire detection and alarm systems can reduce the risk of expensive operational shut downs. This section reviews the factors that should be taken into account when assessing the need for automatic fire detection and provides information about the various types of fire detector. The design of fire detection systems for computer installations is covered in detail in **BS 6266** and only a summary of the main recommendations has been included here.

7.2 The need for automatic fire detection

The need for automatic fire detection equipment depends on the following factors:

* operational staff levels:
  - if the site is continuously manned any outbreak of fire is likely to be detected by staff and automatic detection equipment will not be necessary
  - if the site is operated unattended (some or all of the time) then an automatic fire detection system should be installed

* the risk to the installation: organizations should carry out a risk analysis, using for example, **CRAMM**, to establish measures to be taken.

7.3 Types of detector/detection systems

7.3.1 General

Fire detectors are classified according to the characteristic of fire to which they respond, namely smoke, heat or flame.

7.3.2 Smoke detectors

Smoke detectors give the earliest warning for most types of fire. There are several types of detector, which respond to different characteristics of smoke:

* ionization chamber detectors
* optical point detectors
* optical beam detectors
* aspirating detectors.

* using ducting insulation material which is non-combustible and which should not give off dust or corrosive or toxic fumes when heated

* installing automatic fire dampers (constructed in accordance with **BS 5588: Part 9**) in ducting or trunking that passes through fire compartment walls or floors.

6.3 Air-conditioning plant room

The air-conditioning plant room should be separated from the computer suite by a wall with a minimum fire resistance of one hour. Combustible materials should not be used in the construction of the plant room and oil-filled electrical equipment and other fire hazards including combustible waste should be excluded from it.

6. Air-conditioning

6.1 Introduction

Within the computer suite, the fire risk can be reduced significantly by a suitably designed air-conditioning system. Experience has shown that poorly designed and installed air-conditioning equipment in the computer room presents a higher fire risk than the computer equipment. Important factors in the design of air-conditioning systems are outlined in this section.

6.2 Design of air-conditioning plant and ducting

To minimize the risk of a fire starting within the computer installation, certain features should be designed into the air-conditioning plant and ducting. These features include:

* providing separate air-handling equipment to service the computer installation ensuring that any by-products of a fire elsewhere in the building do not circulate into the computer suite

* using non-combustible materials for casings, acoustic and thermal insulation and filters, and providing thermal cut-outs where self contained air-conditioning units are installed in the computer suite

* providing a positive air pressure in the computer suite to stop infiltration of smoke from a fire elsewhere in the building

* placing the fresh-air intake to reduce the risk of drawing in smoke from any external fire

* providing an extraction system (with manual controls outside the conditioned area) to allow smoke to be removed quickly from the computer suite and for the removal of any extinguishing gases; the extract system can be incorporated in the air-conditioning system

* placing manual controls outside the computer suite to stop air-conditioning fans in the event of a fire inside the conditioned area

* avoiding combustible materials for ducting

Power to the lighting and smoke extraction fans should not be interrupted by emergency switches. The function of these switches should be clearly defined and operating instructions should be provided beside them.

A main isolator controlling all power to the computer suite, except lighting and smoke extraction ventilation, should be provided at or near the main entrance to the suite for use by the fire brigade. This switch, protected by a glass-fronted box, should be marked 'FIRE EMERGENCY SWITCH'.

5.6 Emergency lighting

Suitable emergency lighting should be provided to ensure a safe exit in the event of power failure. **BS 5266:Part 1: Code of Practice for the Emergency Lighting of Premises,** lays down a minimum standard of 0.2 lux, available within 5 seconds of the loss of power.

Section 5
Electrical installations

5. Electrical installations

5.1 Introduction

Apart from arson and accidents, the most likely causes of fire are electrical faults. The risk of such fires can be reduced by ensuring that any electrical installation is to an adequate standard and regular preventive maintenance is undertaken.

5.2 Compliance with IEE Regulations

All electrical work (external to the IT equipment) must comply with the **Regulations for Electrical Installations** issued by the Institution of Electrical Engineers.

5.3 Cables

Power cables, installed in floor and ceiling voids, should be either steel wire armoured or mineral insulated copper covered, or should be contained in metal conduit or trunking. Cable entries to the computer area should be fire-stopped.

5.4 Power outlets

No unnecessary junction boxes should be fitted in the underfloor void. Any junction boxes (for connection to the IT equipment power leads) should be metal, fully enclosed and accessible. If there is a risk of water leakage within the floor void, they should be raised above the slab level. Switched socket outlets (which should be precluded from the floor void) should be fitted with individual indicator lamps.

5.5 Emergency switches

An important safety and fire precaution is the ability to switch off the power supply in the event of an emergency.

Shrouded emergency off-buttons should be sited:

* near the computer control console to cut off all power supplies to the computer equipment

* near exit doors to cut off all power supplies to the equipment, air-conditioning and socket outlets.

one hour and should extend from the structural floor to the soffit of the structural floor or roof above. Doors should be self-closing with a fire rating similar to that of the partitioning. Windows in internal walls should be insulated and able to withstand heat without disintegrating.

Where services or cables pass through the walls or partitions, the holes should be fire-stopped.

4.5 Raised floors

Raised floor should be designed to retain integrity and provide adequate thermal insulation in the event of fire in the underfloor void. Where combustible materials are used, the underside of the raised floor should be covered with non-combustible material.

It is recommended that raised modular floors comply with the appropriate grade specified in the PSA Method of Building publication **PF2.PS Performance Specification: Platform floors**.

Appropriate floor tile lifters should be provided (and be kept in an accessible location).

4.6 Suspended ceilings

Suspended ceilings (and any associated linings) should be of non-combustible materials and, as far as possible, should not shed dust or produce toxic or corrosive fumes while burning. Any material inserted above the ceiling tiles, for acoustic or air balancing purposes, must be non-combustible and contained in sealed bags.

4.7 Storage

Storage of combustible material that is not required for immediate use (for example, more than a day's supply of printer paper), should be in designated areas outside the computer room. Only minimum quantities of flammable cleaning fluids should be kept in the computer area. The main stock should be held in a fire-resistant store remote from the computer area.

Copies of important records (systems and application programs, data files and so on) should be kept in a secure location, in a place which could not be affected by fire in the computer installation (see Section 9.7). Copies of manuals and listings should also be stored away from the suite.

4. Accommodation

4.1 Introduction

Protection is needed against fire risks both from outside and inside the building. Within the building the fire risk can be reduced significantly by separating fire hazards from vulnerable records. Desirable structural precautions are outlined below.

4.2 Sites and buildings

Ideally, a computer installation should be housed in a separate building and not where fire fighting is difficult, for example in basement areas or in the upper storeys of high rise buildings.

The minimum space separation between buildings (required by the Building Regulations) may not be sufficient to protect sensitive equipment from radiated heat, especially if neighbouring buildings were constructed before the Regulations came into force. Structural modifications to external openings and walls may be required, needing professional advice.

4.3 Fire separation

If the computer installation is to be housed in an existing office building, then the separating walls and floors should be constructed of materials of a limited combustibility. **BS 6266** recommends a minimum fire resistance of not less than one hour. If the installation is in a factory-type environment the fire resistance should be not less than two hours. If there is a high fire load in the rest of the building, such as a furniture warehouse, the fire resistance should be not less than four hours.

Rooms below the computer suite should be confined to uses with a low fire risk, such as offices, and not used for the storage of combustible materials. Accommodation on floors above computer suites should be similarly employed to reduce the risk of having to use water to extinguish a fire. The floor above the computer suite should be waterproofed and drainage installed.

4.4 Walls and partitions

Segmentation of operational areas in the computer suite by fire-resistant walls and partitions helps stop the spread of any fire. Partitions should have a fire resistance of at least

3.5.4 PH Government Property/MOD Fire Standards

Fire Standard D6 **Fire Standard D6 - Fire alarm systems, automatically operated** deals with fire alarm systems which give a warning automatically by triggering devices sensitive to the effects of a fire. The Standard summarizes the extensive experience with automatic fire detection (AFD) and points to problems which may arise.

Fire Standard E4 **Fire Standard E4 - Office buildings** deals with structural and engineering fire precautions appropriate to office buildings.

Fire Standard E7 **Fire Standard E7 - Electronic data processing installations** describes the fire precautions to be incorporated in the siting, construction, services and operation of government EDP facilities. In essence, the instruction recommends incorporating the full range of structural and engineering fire precautions set out in **BS 6266**, but not AFD or AFX systems.

3.6 View of fire insurers

Government does not insure buildings or property. However, in the private sector the high value of IT equipment has led insurers to recommend (for a premium rebate) the installation of AFX systems in computer rooms, with no particular regard to technical or financial justification. Insurers are concerned about the cost of paying out on a building and equipment. They do not take into account the cost of spoilt work or downtime due to the unnecessary release of a fire extinguishing gas. In some cases a policy can be invalidated if the AFX system is switched off - even when the building is manned.

The standards applied by many insurance companies for fire fighting equipment are those set by National Fire Prevention Association (NFPA) of America. Many NFPA codes and standards have been adopted as law throughout the USA and are the definitive documents on fire safety throughout the world.

Section 3
Standards and related guidance

- * feedback from AFD systems installed within the government estate gave a general picture of troublesome systems in which clients often lost faith, the main shortcomings being:
    - poor success rate in detecting existing fires
    - lack of specification criteria
    - inadequate commissioning arrangements for installation, acceptance and testing
    - poor system design and workmanship
    - poor reliability
    - high spurious alarm rates
- * money spent on AFD systems for computer installations (that meet the **PSA Technical Instruction F19**) was not, in PSA's opinion, a fair charge to public funds.

Changes since 1982

Since the working party report was published in 1982 a number of changes have occurred which may have a bearing on the level of active fire protection measures installed in computer accommodation.

These changes include:

- * the passing of responsibility for funding (and cost justification) for civil works from PSA to departments
- * the growing dependence of departments on their IT services to conduct their day-to-day business and the generally high availability requirements of those services
- * a move to unattended operating of computer systems
- * technological improvements to fire detection equipment resulting in more effective and more reliable fire protection systems
- * the availability of product and system design standards
- * the availability of comprehensive risk analysis tools, eg the **CCTA Risk Analysis Management Methodology (CRAMM)**, to help departments to assess the need for AFD and AFX systems.

Recommendations

A summary of the report's main recommendations follows:

1. Fire precautions set out in **BS 6266 (1982)** should be implemented as soon as possible, but without retrospective action.

2. Automatic fire detection (AFD) should be provided only where:

    * the client department formally states, as part of the brief to PSA, its intention to run unattended equipment in the computer suite in the normal course of events

    * structural and engineering fire precautions cannot be met.

3. In cases where AFD is installed PSA technical guidance on AFD (**Technical Instruction F19**) should be followed.

4. Automatic fire extinguishing (AFX) equipment should not be installed in government departments unless the client department demonstrates compelling circumstances backed by sound technical criteria; approval of Head of PSA Fire Branch should be sought.

5. Where the decision is taken to install AFX the client department should be asked to accept formally full responsibility for any malfunction or ineffective operation of such equipment.

Those recommendations were based primarily on the following factors:

* there was no evidence in any case histories of the successful use of AFD in the protection of computer equipment or accommodation from fire

* PSA fire protection policy is based on good design, good engineering method and passive structural fire protection

* coupled with disciplined housekeeping by the client department, adequate fire protection by these methods is feasible without resort to AFD and AFX systems

Section 3
Standards and related guidance

Section AM 406 of the Guide sets out the respective responsibilities of PH and the government departments using the buildings within the framework of the **Fire Precautions Act 1971** and the **Fire Safety and Safety of Places of Sport Act 1987**.

The government department using a building is responsible for day-to-day fire precautions and taking immediate action in the event of a fire. Responsibilities are explained in detail in the **PH Fire Precautions Guide**.

3.5.2 PH Fire Precautions Guide

The Guide is divided into three sections.

1. A general section which details the responsibilities of Fire Precautions Officers, and includes information on general fire precautions and fire fighting equipment.

2. A section covering specific areas and subjects, for example, inspection and testing of sprinkler installations and automatic fire detection systems.

3. A set of forms that are filed with the guide and used where appropriate for record purposes.

3.5.3 PSA/CCTA Working Party Report - Published 1982

Background

In the early 1980s there was a significant increase in the number of requests from departments for PSA to protect computer installations with automatic fire detection (AFD) and automatic fire extinguishing (AFX) systems. A PSA/CCTA working party was established to examine the benefits and problems of automatic fire detection and suppression for computer installations and to make recommendations including guidelines for future installations.

The concept that special attention must be given to the protection of mainframe computers has been generally accepted for some time. With the increasing use of IT in offices it is important to be aware of what, if any, additional fire precautions should be taken. Risk factors are assessed and guidance given on the precautions that may be taken in Section 10.

The benefits, costs and possible problems of following the advice detailed in this module are outlined in Section 12.

2.4 Related guidance

This module is one of a series issued as part of the CCTA IT Infrastructure Library. Although the module can be read in isolation, it should be used in conjunction with other IT Infrastructure Library modules.

The following modules are of direct relevance and the reader is referred to these for additional information on specific topics.

The module on **Accommodation Specification** provides guidance on the preparation and content of an Accommodation Design Brief for a computer centre.

The module **Office Design and Planning** provides guidance on the design, planning and furnishing of an office to support IT users and to use space effectively.

The **Contingency Planning** module gives guidance on the plans that are needed to ensure that satisfactory levels of IT service can be provided in the event of a disaster, such as a fire or flood, affecting an organization's IT infrastructure.

The **Computer Operations Management** module gives guidance on how to plan and manage the operation of mainframe computers and related equipment.

The **Unattended Operating** module gives advice on the planning, implementation and management of unattended operating.

2.5 Referenced documents

Documents referred to in this module are listed in Section 13, Bibliography.

2.6 Definition of terms

Terms and acronyms used in this module are defined in Annex A, Glossary of terms.

Section 2
Introduction

2. Introduction

2.1 Purpose

The purpose of the module is to inform organizations about the fire risk to, and fire protection available for, computer rooms and office accommodation where IT equipment is installed. The module gives specific guidance on how to assess the need for automatic fire detection or automatic fire extinguishing systems.

2.2 Target readership

The module is directed at:

* heads of IT services
* IT project teams
* computer operations and network managers
* departmental accommodation officers and installation planners.

2.3 Scope

The module covers the principal aspects of fire precautions in both dedicated computer accommodation and offices containing IT equipment.

A background to the subject of fire precautions is given in Section 3 which covers the purpose of fire protection, standards and regulations applicable, and current CCTA and Property Holdings guidance.

Sections 4 to 9 of the module cover the factors involved in the fire protection of computer rooms. Those factors include:

* siting of the accommodation
* construction materials
* internal layout
* quality of electrical installations
* design of the air conditioning system
* fire detection and extinguishing equipment
* operational fire precautions
* training of personnel.

The IT Infrastructure Library
Fire Precautions in IT Installations

Fire prevention

The installation of fire detection and extinguishing systems will not in itself prevent losses in the case of a fire. The prevention of fire is more important as even a small fire can lead to significant losses.

The key to reducing the risk of fire is sound management of potential problem areas, which include:

* a full appreciation of the value of elementary fire precautions, primarily training, to reduce the probability of an outbreak

* a thorough understanding of the consequences of fire.

It is strongly recommended that the standard of fire protection should be reviewed periodically to ensure that it remains appropriate to the level of perceived risk.

Section 1
Management summary

1. Management summary

Any fire in the vicinity of IT equipment can cause significant damage resulting in loss of data and disruption of the business of organizations. Fires are rare events in both specialized IT accommodation and office buildings. However, the infrequency of such fires does not obviate the need for high standards of fire protection.

Scope

This module covers the principal aspects of fire precautions in dedicated computer accommodation and where IT equipment is installed in office accommodation. Fire precautions include checking out and, if necessary, re-designing building aspects, electrical services, air conditioning and furnishings and fittings, so-called passive measures. Active measures range from every day operational precautions to the provision of fire detection and extinguishing systems.

The module draws upon the latest knowledge of fire protection engineering and discusses how to apply this information to computer centres and the office environment.

Current CCTA and Property Holdings guidance documents have been reviewed together with current legislation, standards and codes of practice on fire protection.

The computer room

Specific recommendations are made for the use of automatic fire detection and extinguishing systems in dedicated computer accommodation. These include:

* installation of automatic fire detection equipment if equipment will be operated in an unattended mode

* installation of automatic fire extinguishing equipment only where justified.

The office

A review of the fire risk in the modern office has shown that:

* the fundamental causes of fire in offices remain similar to those that have existed for some time - but with the increase in IT equipment there are more potential sources of an electrical fire

* with organizations becoming more and more dependent on IT, business disruption and the financial consequences of a fire are likely to be far more damaging for an organization than before.

Each module commences with a **Management summary** aimed at senior managers (Directors of IT and above, typically down to Civil Service Grade 5), Senior IT staff and, in some cases, users or office managers (typically Civil Service Grades 5 to 7). The target readership for the main text is variable and identified in the **Introduction** section of each module. Wherever possible technical detail is confined to annexes.

If you have any comments on this or other modules, do please let us know. A **Comments sheet** is provided with every module. Alternatively you may wish to contact us directly using the reference point given in **Further information**.

Foreword

Welcome to the IT Infrastructure Library module on **Fire Precautions in IT Installations**.

This module is one of a series in the Environmental Sets of the Library. In their respective areas the IT Infrastructure Library publications complement and provide more detail than the IS Guides.

The ethos behind the development of the IT Infrastructure Library is the recognition that organizations are becoming increasingly dependent on IT in order to satisfy their corporate aims and meet their business needs. This growing dependency leads to growing requirement for quality IT services. Quality means 'matched to business needs and user requirements as these evolve'.

The publications forming the major part of the Library are a series of codes of practice intended to facilitate the quality management of IT services and of the IT Infrastructure. (By IT Infrastructure, we mean an organization's computers and networks - hardware, software and computer-related communications, upon which application systems and IT services are built and run). The codes of practice will assist organizations to provide quality IT services in the face of skill shortages, system complexity, rapid change, current and future user requirements, growing user expectations, etc. Details of these modules are available from CCTA Infrastructure Management Services in Gildengate House.

Supporting the IT Infrastructure is the Environmental Infrastructure. It is recognized that environmental issues, from building specification to the practicalities of cable distribution, lighting, noise, power, etc, are less well understood in IT service organizations than the IT and its infrastructure. However, these issues can be just as important in delivering a quality IT service.

The Environmental Sets of modules provide guidance on addressing environmental issues. Their aim is to assist the implementation and management of an environmental infrastructure to support the needs of an organization's IT services.

9.5	End-of-the-day fire precaution routines	36
9.6	Automatic fire extinguishing systems	36
9.7	Contingency measures	37
10.	**Fire precautions in the office**	**39**
10.1	Introduction	39
10.2	BS 5588	39
10.3	The fire risk	39
10.4	Fire protection facilities	41
10.5	Operational fire precautions	42
11.	**Support services**	**45**
11.1	DOE Property Holdings and PSA Services	45
11.2	Health and Safety Executive	45
11.3	Fire Research Station	46
11.4	Loss Prevention Council	46
11.6	Fire Protection Association	46
12.	**Benefits, costs and possible problems**	**47**
12.1	Benefits	47
12.2	Costs	47
12.3	Possible problems	47
13.	**Bibliography**	**49**

Annexes

A.	**Glossary of terms**	**A1**
B.	**The VESDA System**	**B1**
C.	**The STAMP fire detection system**	**C1**
D.	**The HART detector**	**D1**
E.	**Halon - an effective, clean and safe fire extinguishing agent?**	**E1**

5.	Electrical installations	15
5.1	Introduction	15
5.2	Compliance with IEE Regulations	15
5.3	Cables	15
5.4	Power outlets	15
5.5	Emergency switches	15
5.6	Emergency lighting	16
6.	**Air-conditioning**	**17**
6.1	Introduction	17
6.2	Design of air-conditioning plant and ducting	17
6.3	Air-conditioning plant room	18
7.	**Detecting fires**	**19**
7.1	Introduction	19
7.2	The need for automatic fire detection	19
7.3	Types of detector/detection system	19
7.4	Design considerations for automatic fire detection systems	21
7.5	Acceptance testing	23
8.	**Putting out fires**	**25**
8.1	Introduction	25
8.2	Substances used for putting out fires	25
8.3	Portable fire fighting equipment	28
8.4	The need for automatic fire extinguishing systems	28
8.5	Design considerations for automatic fire extinguishing systems	29
8.6	Acceptance testing of gaseous total flooding systems	33
9.	**Operational fire precautions**	**35**
9.1	Introduction	35
9.2	Smoking	35
9.3	Staff training	35
9.4	Fires during working hours : procedures to be followed	35

Table of Contents

	Foreword	vii
1.	**Management summary**	1
2.	**Introduction**	3
2.1	Purpose	3
2.2	Target readership	3
2.3	Scope	3
2.4	Related guidance	4
2.5	Referenced documents	4
2.6	Definition of terms	4
3.	**Standards and related guidance**	5
3.1	Introduction	5
3.2	Legislation	6
3.3	British Standards Codes of Practice	6
3.4	European Standards	8
3.5	Existing CCTA and Property Holdings (PH) guidance	8
3.6	View of fire insurers	12
4.	**Accommodation**	13
4.1	Introduction	13
4.2	Sites and buildings	13
4.3	Fire separation	13
4.4	Walls and partitions	13
4.5	Raised floors	14
4.6	Suspended ceilings	14
4.7	Storage	14

© Copyright: Controller of HMSO, 1991

First published: 1991

ISBN: 0 11 330553 2

This is one of the books published in the IT Infrastructure Library series. At regular intervals, further books will be published and the Library will be completed by early 1992. Since many customers would like to receive the IT Infrastructure Library books automatically on publication, a standing order service has been set up. For further details on standing orders please contact:

HMSO Publicity (P9D), FREEPOST,
Norwich, NR3 1BR
(*No stamp needed for UK customers*).

Until the whole Library is published, and subject to availability, draft copies of unpublished books may be obtained from CCTA if you are a standing order customer. To obtain drafts please contact:

Environmental Infrastructure Services
CCTA
Riverwalk House, 157-161 Millbank
London SW1P 4RT.

For further information on other CCTA products, contact

Press and Public Relations,
CCTA
Riverwalk House, 157-161 Millbank
London SW1P 4RT.

This document has been produced using procedures conforming to
BSI 5750 Part 1: 1987; ISO 9001: 1987.

Fire Precautions in IT Installations

IT Infrastructure Library

Chris Kiddle

Riverwalk House,
157-161 Millbank
London, SW1P 4RT

LONDON: HMSO